FRA

D0899932

Eutrophication of Freshwaters

Principles, problems and restoration

DAVID HARPER

Ecology Unit
Department of Zoology
University of Leicester

With a chapter on the Norfolk Broads by Geoff Phillips

CHAPMAN & HALL
London · New York · Tokyo · Melbourne · Madras

Published by Chapman & Hall, 2-6 Boundary Row, London SE1 8HN

Chapman & Hall, 2-6 Boundary Row, London SE1 8HN, UK

Chapman & Hall, 29 West 35th Street, New York NY10001, USA

Chapman & Hall Japan, Thomson Publishing Japan, Hirakawacho Nemoto Building, 7F, 1-7-11 Hirakawa-cho, Chiyoda-ku, Tokyo 102, Japan

Chapman & Hall Australia, Thomas Nelson Australia, 102 Dodds Street, South Melbourne, Victoria 3205, Australia

Chapman & Hall India, R. Seshadri, 32 Second Main Road, CIT East, Madras 600 035, India

First edition 1992

© 1992 Chapman & Hall

Typeset in 11 on 12½pt Sabon by
Columns Design and Production Services Ltd
Printed in Great Britain by
St Edmundsbury Press, Bury St Edmunds, Suffolk

ISBN 0 412 32970 0

Apart from any fair dealing for the purposes of research or private study, or criticism or review, as permitted under the UK Copyright Designs and Patents Act, 1988, this publication may not be reproduced, stored, or transmitted, in any form or by any means, without the prior permission in writing of the publishers, or in the case of reprographic reproduction only in accordance with the terms of the licences issued by the Copyright Licensing Agency in the UK, or in accordance with the terms of licences issued by the appropriate Reproduction Rights Organization outside the UK. Enquiries concerning reproduction outside the terms stated here should be sent to the publishers at the London address printed on this page.
   The publisher makes no representation, express or implied, with regard to the accuracy of the information contained in this book and cannot accept any legal responsibility or liability for any errors or omissions that may be made.

A catalogue record for this book is available from the British Library

Library of Congress Cataloging-in-Publication data available

QH
96.8
.E9
H37
1992

Contents

Preface

Eutrophication is a problem which became widely recognised by the scientific community in the 1940s and 1950s. It raised public concern, resulting in increased research effort and expenditure on management techniques through the 1960s and 1970s, recognised as a distinct problem of water pollution, though linked with the more gross effects of organic pollution. In the 1980s it became less fashionable – replaced in the public's eye and the politician's purse by newer problems such as acid rain. It remains however, one of the biggest and most widespread problems of fresh waters, particularly of lakes and an increasing problem for estuaries and coastal waters. It is one with which almost all water scientists and engineers in urbanised areas of the world have to cope. Technical methods for the reversal of eutrophication, such as nutrient removal, have been developed and applied successfully in some instances. They are not widespread however, and where they are feasible, they are often expensive and may be politically difficult to implement. In the last decade, attention has focussed upon less expensive lake manipulation techniques, such as destratification and biomanipulation, which aim to minimise rather than elimininate the detrimental effects of eutrophication. These are becoming more widely applied. Prediction of the potential problems in lakes and catchments which have not yet suffered the full effects of eutrophication is now accurate enough to be of direct benefit to river basin management.

Wide interest in the subject area has generated a large number of research publications and from the late 1960s onwards, over a dozen major international conference proceedings. Notable among the latter are Eutrophication – Causes, Consequences, Correctives; U.S National Academy of Sciences (Rohlich, 1969), Nutrients and Eutrophication; American Society of Limnology and Oceanography (Likens, 1972), Lake Restoration; U.S. Environmental Protection Agency (USEPA, 1979), Hypereutrophic Ecosystems; International Association of Limnology (Barica, 1980), Phosphorus in Freshwater Ecosystems; Swedish National Environmental Protection Board

(Persson, 1988). Additionally, there have been many national conferences in the subject area, including those which have brought together the results of intensive studies of individual lakes whose ecology and management were investigated as part of the International Biological Programme 1964-1974 (Worthington, 1975), such as Loch Leven Scotland (Anon., 1974). The international collaboration over IBP was parallelled and extended by developments sponsored by the Organisation for Economic Co-operation and Development (OECD), notably the initial 'state of the art' review by Vollenweider (1968), subsequent exploration of the quantitative links between nutrient supply and biological response of lakes (Anon., 1982) and a more detailed analysis of water pollution by fertilisers and pesticides (Anon., 1986).

There are also several specialist books and reports on eutrophication; its causes, effects and management. Early notable ones were Nutrients in Natural Waters (Allen and Kramer, 1972) Nitrogen and Phosphorus (Porter, 1975) and Comprehensive Management of Phosphorus Water Pollution (Porcella and Bishop, 1976). In the last few years three more have appeared: Lake and Reservoir Restoration (Cooke *et al.*, 1986), Lake Restoration by Reduction of Nutrient Loading (Sas, 1989) and The Control of Eutrophication of Lakes and Reservoirs (Ryding and Rast, 1989). The outlines of eutrophication are covered in most limnological textbooks, particularly Limnology (Goldman and Horne, 1983), The Ecology of Freshwaters (Moss, 1988), Ecological Effects of Waste Water (Welch, 1980) and Biology of Freshwater Pollution (Mason, 1981). It is a little surprising however, that there has been no widely available introductory textbook on eutrophication, Decaying Lakes (Henderson-Sellars and Markland, 1987), being somewhat expensive and with a technical bias. Perhaps this is because the effects of eutrophication are felt at every level of the freshwater ecosystem and the challenge of understanding eutrophication is the challenge of understanding limnology as a whole. I feel that there is a need for a textbook which introduces the whole field of eutrophication to the working water scientist or engineer, who may be trained in a single discipline and to the student developing an interest in applied limnology anywhere in the world. I hope that this book fulfills that need.

Many people contributed along the way to the completion of this book. My tutors at Oxford in the early seventies, John Lawton, John Phillipson, Malcolm Coe and John Anderson, having taught me terrestrial ecology, helped me find a postgraduate place even when I announced an interest in aquatic ecology. Bill Stewart, my postgraduate supervisor at Dundee, and Hugh Ingram, first interested me in eutrophication of freshwaters when I was starting a Ph.D. on mercury pollution in estuaries. Colleagues in Anglian and Severn-Trent Water Authorities, particularly Geoff Phillips, John Hellawell, Alastair Ferguson and Peter Barham, helped me to learn the practical problems of eutrophication management between 1975 and 1979, and continued to work with me after I returned to academia. My first three postgraduate students and friends at Leicester; Bill Brierley, Colin Smith and Andy Smart, helped shape my opinions and ideas, didn't laugh too much when I said I was going to write a book (or when the deadlines slipped), helped with literature and contributed original work to it. My Masters' students in 1989 and 1990 produced reviews which helped keep me abreast of the literature as my teaching and administration threatened to sink me without trace. Friends who have worked with me at Lake Naivasha over the past few years, particularly Bill, Andy, Geoff, Frank Clark, Muchai Muchiri, Rick North, Phil Hickley, and Maureen Parsons, gave me inspiration and support which lasted well beyond the few weeks of field research. My wife Kathryn, Geoff, Phil and Maureen proof-read the manuscript in its early stages. I am particularly grateful to Geoff for agreeing to write Chapter 8 on the problems of the Norfolk Broads, to Lynnda Aucott who drew all the figures, to Pete Garfoot who checked the references and indexed the text, to the manufacturers of Benson and Hedges and Apple Macintosh which kept me going and Glenmorangie which helped me wind down. My largest debt is owed to Kathryn and our children Shona and Christopher, who allowed me the time to write. Finally, I thank my numerous colleagues and students at Leicester for their patience when I was never in the right place at the right time. I doubt if I ever will be.

— 1

What is eutrophication?

1.1 INTRODUCTION

'On a clear, sunny day, the Potamogetae, flourishing at a great
depth amid the transparent waters, animated by numerous
members of the insect and finny races present a delightful
spectacle, and the long stems of the white and yellow water
lilies may be traced from their floating flowers to the root.
. . . . .One feels in such a place estranged for a time from the
cares and vicissitudes of the world and the charms of nature
penetrate, with their refining influences, the deepest recesses of
the heart, denying to human language the power to give them
full expression.' W. Gardiner, 1848. The Flora of Forfarshire.

This quotation, nearly a hundred and fifty years old, made about
a shallow lake in Scotland, Balgavies Loch, eloquently summarises
human beings' affinity for lakes. It describes however, a loch whose
mean transparency measured with a white 'Secchi' disc is less than a
metre as a result of dense growths of planktonic algae throughout
the summer. The submerged plants grow no deeper than 2 metres
and in the 1970s included just three species of *Potamogeton*, where
once there were 17 species or hybrids (Harper, 1986).

This book is about the sequence of events which have caused
such differences to develop in lakes, reservoirs and rivers all over
the world. Moreover, it is about the explanations for these events
and our attempts to manage their worst effects. The effects which
cause greatest concern are losses of water use and amenity – such as
increased problems of potable water treatment, detrimental changes
in fisheries – which have an economic base. There are also less
tangible effects, such as a reduction in species diversity and amenity
value of affected water bodies.

1.2 DEFINITION AND ORIGIN OF THE TERM EUTROPHICATION.

Eutrophication, or nutrient enrichment and its effects, is generally seen as a problem of the middle and late stages of the 20th century. In this context it is artificial eutrophication; a consequence of society's urban, industrial and agricultural use of plant nutrients and their subsequent disposal. Eutrophication is, however not only a man-made problem because any changes within a catchment, natural or otherwise, will influence the biological state of its lakes and rivers. Moreover, it is not new; in some parts of the world man has exploited the benefits of artificial enrichment through enhanced fish yields from ponds and lakes for many centuries. What is new in the 20th century is the extent of enrichment of lakes and rivers throughout the world and, until recently, our relative lack of control over the sources of the nutrients or over their effects upon the aquatic ecosystem.

Eutrophication is the term used to describe the biological effects of an increase in concentration of plant nutrients – usually nitrogen and phosphorus, but sometimes others such as silicon, potassium, calcium, iron or manganese – on aquatic ecosystems. It is difficult to define precisely, because a description of the trophic nature of any one lake, river or estuary, is usually made relative to a previous condition, or to a reference state of lower nutrient concentration, called mesotrophic (intermediate) or oligotrophic (low in nutrients).

This lack of precision in the definition is partly due to the individual nature of every body of water in its response to nutrients, and partly due to the introduction and early use of the term. The adjective eutrophe was first used by the German botanist, Weber (1907) to describe the nutrient conditions which determine the plant community in the initial stages of development of raised peat bogs. These begin life rich in nutrients, but as organic matter accumulates and the bog grows upwards, nutrients become leached out by rainfall. Weber described the nutrient stages which subsequently control the vegetation changes as leading from 'eutrophe' to 'mesotrophe' and then to 'oligotrophe'.

The three terms were used in limnology some twelve years later by the Swedish botanist Naumann, to describe freshwater lake types which contained low, moderate or high concentrations of phosphorus, nitrogen and calcium (Naumann, 1919). The concentrations in each category were not stated; rather, Naumann categorised their effects upon lake phytoplankton (the free-floating algae) and transparency. Thus an oligotrophic lake was clear, blue and contained little phytoplankton; a eutrophic lake was more turbid and green from dense phytoplankton growth. A mesotrophic lake was intermediate between the two.

These concepts fitted well with other contemporary approaches to the classification of lakes which were being developed in Europe at the time. Teiling (1916), following the work of West and West (1909), developed a classification of European lakes by their taxa of dominant phytoplankton. The two extremes were the deep highland lakes in Britain and Scandinavia which were dominated by desmids, species of green algae, and the shallow lakes of the Baltic, dominated by species of blue-green algae, now more correctly called cyanobacteria. Thienemann (1918) classified lakes in the German Eifel district on the basis of their seasonal oxygen regime and differences in their benthic (bottom-dwelling) fauna. He found that in deeper, unproductive lakes whose bottom waters were not depleted of oxygen in the summer months as a result of the decay of organic matter raining from the upper layers, a diverse fauna existed. This contrasted with shallower lakes whose organic matter production during the summer was great. Decay of this organic matter caused the lower layers to experience oxygen depletion. The fauna of these lakes characteristically consisted of a few species tolerant of low oxygen conditions. Around the same time, in England, Pearsall (1921, 1932), classified the lakes of the English Lake District using their species of macrophytes (rooted aquatic plants) and phytoplankton. He compared the plant assemblages with features of the drainage basin, such as the percentage of the drainage area which was cultivable and percentage of the lake bottom to 9 m which was rocky. He used the terms 'evolved' and 'unevolved' (which would be roughly comparable with eutrophic and oligotrophic) to describe the extent of erosion in the catchment since the lake's formation which had brought silt and nutrients into the lake, giving rise to characteristic plant associations. Figure 1.1 shows the outline differences between eutrophic and oligotrophic lakes.

Working on the same series of lakes, Mortimer (1941, 1942) advanced our understanding of the evolutionary changes in lakes with an explanation of the processes which control the exchange of dissolved substances between lake water and sediment. He found that the degree of oxidation–reduction at the sediment surface, as a consequence of deoxygenation of the overlying water, determines the movement of solutes between sediment and water (of which phosphorus is the most important). If the mud surface is oxidised,

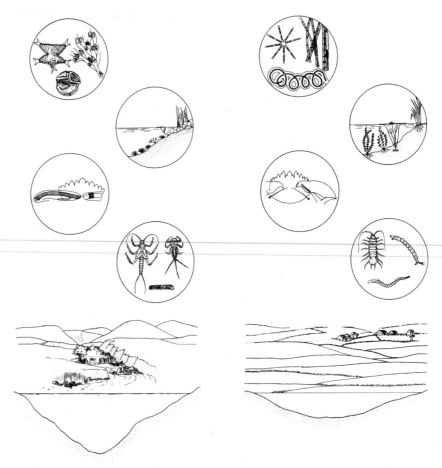

Figure 1.1 Diagrammatic representation of some of the differences between a hypothetical oligotrophic upland lake (left) and a eutrophic lowland lake (right). Examples of indicator organisms or their remains are shown in circles. From top to bottom they are phytoplanktonic algae; rooted littoral (shore) plants; sediment microfossils (chironomid larval mouthparts); and littoral invertebrates.

iron is present in the ferric state precipitated in a complex colloidal structure which prevents phosphate exchange. If the mud surface becomes anaerobic however, ferric is reduced to ferrous, the complexes break down and phosphate moves from the sediments into the overlying water. Mortimer distinguished three stages in lake evolution. Stage 1 is a long period when the surface mud remains permanently oxidised and the sediments accumulate

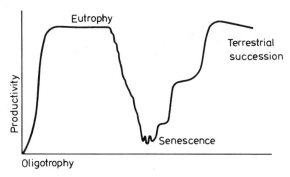

Figure 1.2 The concept of lake productivity change with age from oligotrophy to eutrophy. Modified from Lindeman (1942), with permission.

nutrients. Stage 2 is a period when the mud surface becomes reduced, at first intermittently, releasing nutrients and other ions back into the water body. Stage 3 is a period of permanent deoxygenation, where the iron is precipitated as sulphide.

Ideas of lake evolution were then taken further in the United States, where more quantitative and dynamic ideas of ecosystems were developing in terms of their energy transformations. Lindeman (1942), in a now classic paper which developed the concepts of energy flow through the different levels of aquatic food webs, suggested that eutrophication was a natural stage in a lake's life as it gradually filled in with sediment eroded from its catchment and with organic matter from its own metabolism. Starting from an oligotrophic stage which had low productivity, a typical temperate lake in glaciated regions would have increased in productivity fairly quickly as nutrients accumulated, until it reached a steady state of eutrophy (Fig. 1.2). This might last for a long time – thousands of years – until the lake became too shallow for effective phytoplankton growth or regeneration of nutrients, and developed into a closed swamp or fen (Fig. 1.3).

1.3 LINKS BETWEEN EUTROPHICATION, BIOLOGICAL CHANGES AND PRODUCTIVITY IN LAKES

The supply of nutrients is one of the most important factors which determines the species and quantity of plant material in lakes, which in their turn control the oxygen concentrations and the animal species. Details of these interrelations will be explored fully

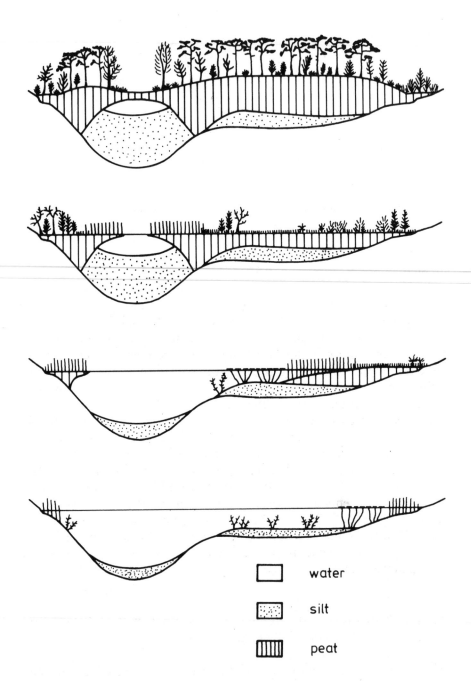

	water
	silt
	peat

Figure 1.3 Diagrammatic representation of the successional changes of a lake from open water to closed fen. Modified from Sinker (1962), with permission.

in succeeding chapters, but an introduction to the biological processes associated with nutrient changes is necessary here.

1.3.1 Direct and indirect biological changes

The biological changes which occur in lakes as a result of eutrophication can be separated into those which are the direct result of raised nutrient influx, such as the stimulation of algal growth, and those which are the indirect effect, such as changes in the fish community as a result of reduced oxygen concentrations.

The direct effects occur when organisms, usually planktonic algae, are released from nutrient-limited growth. In any ecosystem, the transformation of solar energy into the production of organic matter will rarely be at the maximum physiological capacities of the individual plants living there. For each species it will be limited by the environmental resource in shortest supply at the time, even if others are present in excess of an individual organism's needs. The most usual environmental resources which limit production are light and nutrient (or food in the case of animals) supply. Whichever falls below the minimum level to sustain growth will regulate the population of that species (Hutchinson, 1973). This is often called the Law of the Minimum and is an important ecological concept (Odum, 1971), even though it is a simplistic view of what actually happens. There may be limitation to the rate of growth rather than to the upper limit of population biomass (Fig. 1.4). In lakes the limiting factors are usually phosphorus and light for algae; food availability and temperature for animals.

There are also indirect biological effects of release of a population from the state of limitation by an adequate supply of all resources. Growth rate increases and may be such that the population then comes into competition for one or more other resources with neighbouring species. One consequence of this may be the replacement of a lesser competitor by another which is more efficient in its use of the resource. Thus, with an increase in phosphorus supply to a lake, algal biomass will increase and the species contributing to the biomass may change.

Other indirect effects of eutrophication can occur when an increase in production of any one species' population has effects upon the physico-chemical environment (such as the oxygen concentration as a result of a bacterial population's respiratory uptake or the light quantity as a result of an algal population's density) in which it lives. This will affect other species sharing that environment but not directly competing for resources within it.

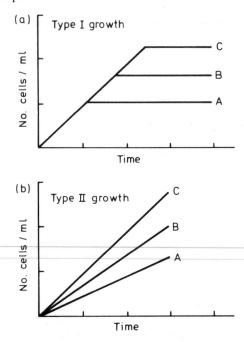

Figure 1.4 Graphical representation of the difference between the effects of a limiting factor on biomass achieved (upper graph) and on growth rate (lower graph). A, B and C represent different concentrations of a limited resource. Modified from O'Brien (1972), with permission.

1.3.2 The sequence of changes with eutrophication

The sequence of changes which usually occurs in lakes subject to eutrophication is thus both a direct and indirect consequence of an increase in the concentration of nutrients flowing through each component of the lake ecosystem (Fig. 1.5). The biomass of the components and some of the pathways of flux are increased (Fig. 1.6). An outline sequence is as follows, with an example shown in Table 1.1.

Initially an increase in the influx of salts of nitrogen and phosphorus can be detected, sometimes together with other ions such as those of calcium, potassium, iron, manganese, sulphates and chlorides, depending upon the source (Beeton, 1965, 1969; Owens and Wood, 1968).

These nutrients are taken up by both epiphytic (attached) and planktonic algae. This results in an increase in mean biomass and productivity (Lund, 1969) and changes in their seasonal patterns

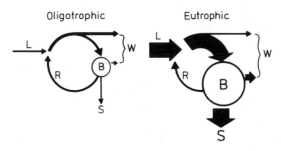

Figure 1.6 Diagrammatic effect of nutrient increase in a lake. Modified from Reynolds (1984a), with permission. L= loading, B= biomass, R= recycling, S= sedimentation, W = washout.

(Rodhe, 1958, 1969). Species changes occur, leading for example in the plankton to fewer species overall and dominance by species of diatom, cyanobacteria and unicellular green algae (Brook, 1964). Smaller shifts may also occur, such as changes in species of diatom within a genus.

Submerged macrophytes may also increase in biomass (Harper, 1986), particularly in calcareous waters (Spence, 1967) but more often macrophyte biomass is reduced with increasing enrichment as a result of competition for light with phytoplankton or epiphytes (Jupp and Spence, 1977) (Phillips, *et al.* 1978). There is also a decline in diversity as species intolerant of low light, higher dissolved solids, or competition, disappear (Seddon, 1972; Spence, 1964).

Changes occur in the oxygen regime of the water body as a result of the accumulation and decay of plant debris from phytoplankton and macrophytes in the bottom muds of the lake. The extent of these changes depends very much upon the shape of the lake basin, particularly its depth. Most lakes over 10 m or so depth are potentially able to stratify. This means that under calm, warm conditions, the upper layer of water a few metres thick accumulates heat and becomes lighter; it floats on the cooler, denser bottom water (Fig. 1.7). This may occur on a daily basis, as in many tropical lakes, or on a seasonal basis in the summer months of the temperate regions. When stratification persists for longer than a few days the layers gradually become chemically and biologically distinct. The upper layer, called the epilimnion, which is illuminated, encompasses most of the phytoplankton primary producers and the zooplanktonic animals which graze on them. It remains

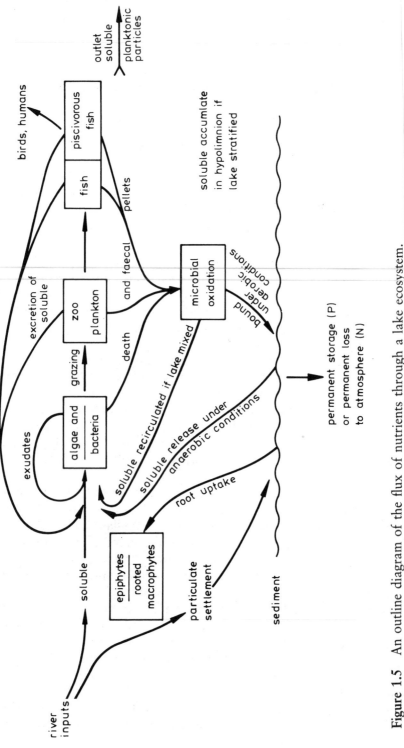

Figure 1.5 An outline diagram of the flux of nutrients through a lake ecosystem.

Table 1.1 An illustration of some of the consequences of eutrophication seen by a comparison of a chain of three neighbouring shallow Scottish lakes. From Harper (1978)

Limnological parameter	Lowes	Lake Name Balgavies	Forfar
Trophic status	Oligo/mesotrophic	Eutrophic	Polytrophic
Catchment land use	Forestry/heathland	Arable	Arable/urban
Conductivity, µS/cm	64	213	439
Winter soluble nitrogen concentration, µg/lN	194	2690	5870
Winter soluble phosphorus concentration, µg/lP	35	77	2460
Summer chlorophyll 'a' concentration, µg/l	8	32	99
Summer Secchi disc depth, m	2.6	1.1	1.2
Maximum summer zooplankton density, numbers/l	35	115	350
Summer benthic invertebrate density, numbers/m^2	13000	22500	27000
Number of benthic invertebrate taxa	81	69	33
Summer macrophyte biomass, g dwt/m^2	44	88	200
Macrophyte colonisation depth, m	4.5	1.5	2
Number of macrophyte species	15	10	4
Dominant fish species	Trout, Perch	Perch, Trout	Stickleback

saturated with oxygen as a result of the activities of the primary producers and diffusion from the atmosphere, but it becomes depleted of nutrients as a result of their incorporation into plant and animal tissue. Dead organisms and faeces sink into the lower layer, called the hypolimnion, where they are decomposed by bacteria both in the water column and the sediments, and provide food for sediment-dwelling invertebrates. Oxygen is depleted here as a result of respiration and the absence of any replenishment from the surface (Fig. 1.8). Some soluble nutrients such as phosphates and soluble ions such as ferrous iron and manganous manganese are liberated by anoxic bacterial activity and accumulate since there is no uptake by primary producers or mixing.

In most lakes the stratification is destroyed by nightime (tropical) or autumnal (temperate) winds which overturn and mix the lake. A few deep lakes have permanent stratification and in these the

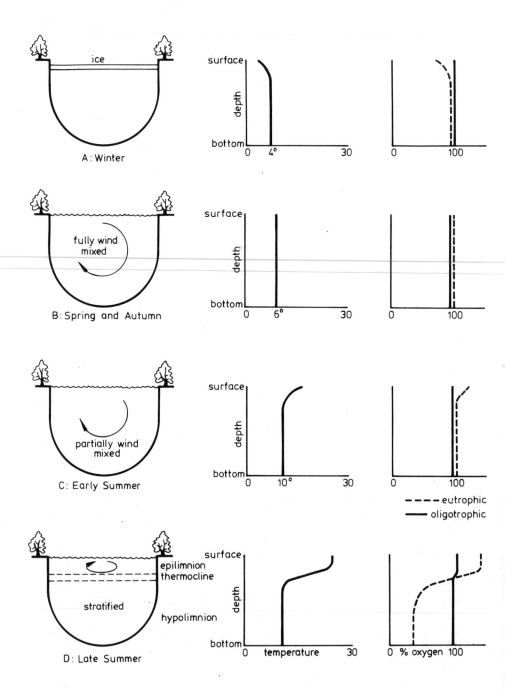

Figure 1.7 The process of stratification in lakes and consequent oxygen depletion in both oligotrophic and eutrophic ones.

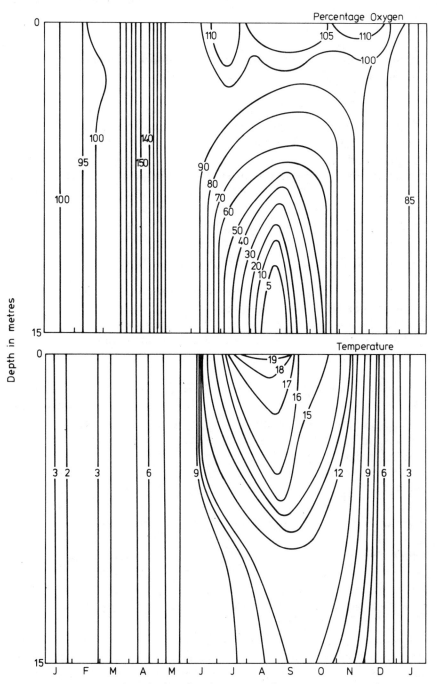

Figure 1.8 Example of the development of temperature and oxygen stratification in a shallow lake, the Loch of the Lowes, Scotland. From Harper (1978).

hypolimnia are permanent, anoxic, nutrient sinks.

The maximum extent of oxygen depletion depends upon the interaction of the following four factors:

1. The volume of the hypolimnion. A larger hypolimnion holds a greater mass of oxygen so that oxygen depletion will occur more slowly than in a hypolimnion of smaller volume.
2. The depth of the lake. The hypolimnion of a deeper lake will remain at a lower temperature than that of a shallower lake of equal area because less heat per unit volume can be conducted from the epilimnion; as a consequence community respiration will be less and oxygen depletion slower. A deeper lake will also have a hypolimnion of greater volume.
3. The quantity of organic debris falling from the epilimnion. This increases as production in the epilimnion increases; hence community metabolism in the hypolimnion and subsequent rate of oxygen depletion will increase.
4. The length of time for which stratification is maintained. Differences such as a sheltered location or a long hot summer, will prolong stratification and enhance oxygen depletion.

Bacterial biomass and production rises as a result of an increase of nutrients (which bacteria may absorb directly in competition with algae) and of organic detritus. Respiration of the bacterial biomass in the hypolimnial water column and sediment is the most important factor removing oxygen.

Both bacteria and algae are consumed by grazing herbivores/ detritivores in the plankton and the sediments. An increase in zooplankton density follows an increase in algal, bacterial and detrital biomass. Changes in species composition may occur, however, both as a consequence of the different efficiencies with which zooplankton utilise food particles of different sizes (larger-bodied forms graze more efficiently) (McNaught, 1975) and through the size-selective impact of fish predation (larger-bodied forms are more susceptible to visually-feeding fish) (Hrbacek *et al.*, 1961; Brooks, 1969).

Qualitative and quantitative changes occur in the bottom fauna as a result of the changes in the oxygen regime and in their food supply from the epilimnion (Deevey, 1942; Jónasson, 1969). Species less tolerant of reduced oxygen levels decline and those that remain, dominated by species of chironomid midge larvae and oligochaete worms (often with adaptations such as haemoglobin in their blood)

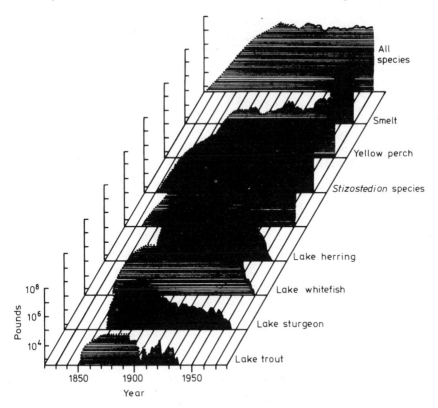

Figure 1.9 The relatively constant yield of fish from Lake Erie over the past century, compared with the change in species composition. Modified from Regier and Hartman (1973), with permission.

may reach very high densities.

Fish species composition may change, primarily as a result of reduced oxygen tension in the hypolimnion which forces oxygen-sensitive species into the warmer, upper layers of the lake. This may lead to thermal stress in cold-water fish such as salmonids and their eventual replacement by species tolerant of higher temperatures and reduced oxygen, such as cyprinids. Fish biomass however, may often increase or at least be maintained as a consequence of increased food supply, even though the change in species may lower the economic value of the fish stock (Fig. 1.9) (Brooks, 1969; Larkin and Northcote, 1969; Regier and Hartman, 1973).

The undecomposed, hard parts of many species of aquatic animals and plants accumulate in the sediments of lakes, offering a potential historical record of the processes of change. Suitable

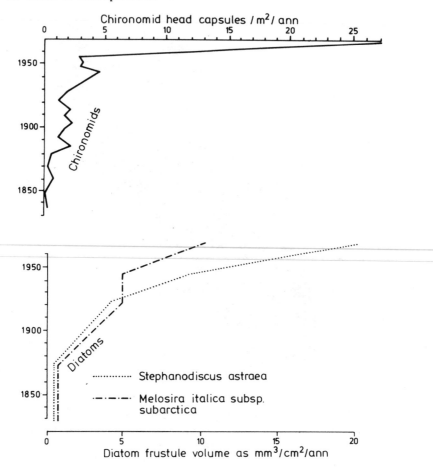

Figure 1.10 An example of the accumulation of algal and invertebrate sedimentary remains in Lough Neagh, Northern Ireland, over the last 150 years, showing the effects of progressive eutrophication. Modified from Oldfield (1977), with permission.

remains are the siliceous frustules (shells) of diatoms and certain other algal groups (Livingstone, 1984), 'pseudofossils' of invertebrate animals such as cladocera (carapace remains) and chironomids (head capsule remains) (Deevey, 1942; Murray, 1979) (Fig. 1.10) and fish scales. These, together with the mineral sediments and terrestrial pollen grains of the catchment and a variety of dating techniques, allow very accurate pictures to be built up from cores of a lake's past history (Oldfield, 1977).

1.4 EUTROPHICATION AS A NATURAL PART OF LAKE SUCCESSION

In order to manage and understand eutrophication it is necessary to distinguish between the natural effects of in-filling and succession slowly affecting lake morphometry, and the effects of accelerated nutrient supply from the catchment. This is important because many lakes and reservoirs now showing adverse effects of enrichment have no history, or little known history. If one can appreciate the likely extent of natural eutrophication then a realistic approach can be developed to the problems of artificial eutrophication.

All lakes and reservoirs have a finite life span; sooner or later they will fill with sediment and, if left alone, be replaced by terrestrial communities. This span may vary though, from a few years for lakes created by the channel changes of rivers, to millions of years for deep large lakes created by movements of the earth's crust. The major consequence of the method of formation, lake depth, is important because in general, the deeper the lake the longer it will take to fill in and because depth may control productivity. The different ways in which lakes are formed, which determines their likely life-span, are well covered in standard textbooks of limnology (Hutchinson, 1957; Moss, 1988; Goldman and Horne, 1983).

Lindeman's idea of succession in a hypothetical temperate lake started with a state of low nutrients, low productivity, and a fully oxygenated hypolimnion – oligotrophic. This gave way gradually to a higher nutrient input and increased productivity which caused greater deoxygenation of the hypolimnion as a consequence of a heavier rain of organic detritus, on a volume progressively reduced by sedimentation both from the detritus and erosion from the catchment. In the European lakes studied by Thienemann and others, unproductive upland lakes were deep and received few nutrients from the catchment. They had permanently oxygenated hypolimnia and were oligotrophic. In contrast shallow lowland lakes had greater input of nutrients from their catchments, productive epilimnia, smaller shallower hypolimnia experiencing oxygen depletion and were eutrophic.

Investigations of lakes in Connecticut, USA by Deevey (Deevey, 1942) showed that the sequence postulated by Lindeman in the aquatic stages of lake succession did indeed occur. Remains of the head capsules of chironomid midge larvae in sediments could be identified to species and classified as indicators of lake type on the basis of their present day distribution. Oligotrophic indicators preceded eutrophic indicators in the lake sediments, even in relatively shallow lakes of around 11 metres, showing that their early history was indeed one of gradually increasing productivity.

Other investigations though, have shown that lakes may pass through periods in their existence where they become less productive and more oligotrophic. For example, in many of the English Lake District lakes formed after the retreat of the last ice-age (Macan, 1970), examination of diatom remains (Haworth, 1969), complemented by chemical analysis (Mackereth, 1966), microfossil and pollen analysis (Pennington, 1981) has indicated initial periods of higher productivity after formation of the lakes as unvegetated catchments were eroded bringing nutrients into the basins. This was followed by stable periods several thousand years long of lower production, with oligotrophic indicator species, as the catchment became vegetated and erosion was reduced. This eutrophic early phase, followed by a stable oligotrophic phase is now recognised to be the more typical natural pattern of north temperate lake succession (Whiteside, 1983). Lakes with an older history, or those subjected to more variable climatic regimes, have often experienced several reverses in productivity. Lake Biwa, an ancient lake in Japan believed to be some 4 million years old, has passed through two oligotrophic phases in the last half million years, interspersed with two mesotrophic and one eutrophic phase (Horie, 1981) (Fig. 1.11). Lake Naivasha, an equatorial Rift Valley lake in Kenya has experienced fluctuating water levels over the past 3000 years as a result of climatic variations, which have made succession in the context of Lindeman's hypothetical lake cease (Richardson and Richardson, 1972).

Earlier periods of natural change in lakes provide a yardstick against which to measure the advance of artificial eutrophication. In the English Lakes the oligotrophic period began to change with the first human colonisers. Productivity and sediment input increased in some but not all lakes coincident with vegetation clearance in the catchment by neolithic humans some 5000 years ago and more widespread deforestation some 2000 years ago. It further increased in the last five centuries with improved methods of cultivation and more human settlement (Haworth, 1985) and has shown its greatest increase, also characterised by elevated levels of carbon, nitrogen and phosphorus, since 1930 (Pennington, 1981) (Figs. 1.12, 1.13).

Sedimentary evidence of early, man-induced changes may also

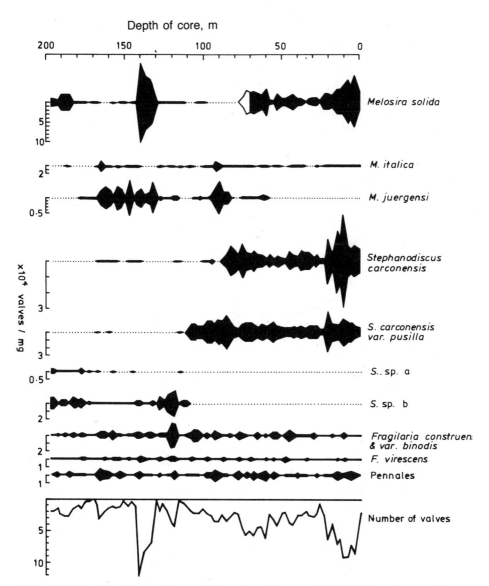

Figure 1.11 The fluctuating periods of enrichment in Lake Biwa shown in the diatom record of its sediments. Modified from Horie (1981), with permission.

Productivity

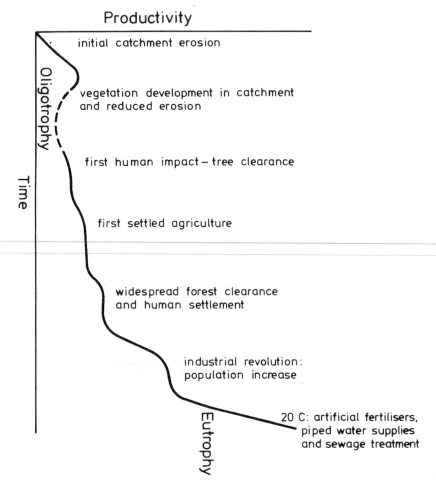

Figure 1.12 An extension to the lake phase of Lindeman's successional graph (Fig. 1.2) to show the more usual pattern of north temperate lake post-glacial changes.

provide a yardstick against which to measure 20th century changes. A well-documented single effect is the building by the Romans of a road, the Via Cassia, across the catchment of a small Italian lake, Lago di Monterosi, around 170 years BC. Hutchinson and Cowgill (1970) have shown that the rate of sediment and of nitrogen deposition, stable for some 24,000 years and indicating an unproductive lake, increased suddenly at the time of the road, probably as a result of woodland clearance and erosion. The

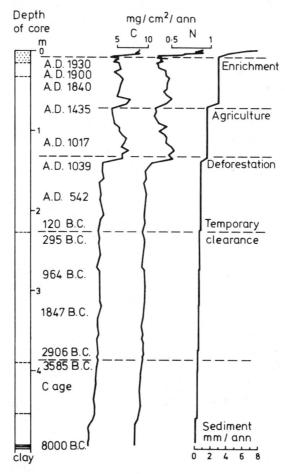

Figure 1.13 Changes in the sediments of Blelham Tarn, in the English Lake District together with radiocarbon dates and human activities. Modified from Pennington (1981), with permission.

increased productivity which resulted existed for around 600 years before declining again. A similar explanation has been advanced for a dramatic change in stratigraphy of cores from Llangorse Lake, South Wales, which peaked about 100 years after the Roman defeat of Silurian tribes and the consequent expansion of roads and garrisons (Jones *et al.*, 1978).

Initial conversion of natural vegetation into agriculture took place at different times across the globe but had similar effects upon their catchments. Most English lakes, e.g. in the Lake District and elsewhere, such as Crose Mere, in Shropshire (Reynolds, 1979) show changes in their sedimentary diatom and cladoceran remains which indicate the start of enrichment coincident with the onset of settled agriculture several centuries ago (as evidenced by increases in the pollen of agricultural weeds and archaeological finds in the lakes' vicinities). North American lakes showed similar changes more recently: Gull Lake, Michigan began to show changes at about the time of European settler colonisation of the area about a hundred years ago. Most rapid diatom changes occurred in the upper part of the sediment examined, coincident with the post 1940 period of rapid settlement of the catchment (Moss, 1972). In Lake Washington, stable conditions lasting several thousands of years were indicated by the diatoms *Melosira italica* and *Stephanodicus astraea* but the more eutrophic indicator *Fragilaria crotonensis* became dominant at the time when the town of Seattle began to grow on the shoreline (Stockner and Benson, 1967). Even some man-made lakes of sufficient age, such as the Norfolk Broads, England (which are a series of medieval peat diggings), show historical changes which can be compared with contemporary ones. In Alderfen Broad for example, sedimentation and diatom deposition increased between 1854 and 1909 with an increase in the number of domestic buildings in the catchment area. This early increase was mainly represented by epiphytic (attached littoral) diatoms; planktonic diatoms began to increase after the 1940s coincident with a further phase of house building (Moss *et al.* 1979a).

Oldfield (1977) showed how changing mineral composition in most British lakes could be compared with very different patterns in a New Guinea highland lake where lead-dating techniques and mineral evidence combined with pollen analysis showed very stable low rates of sedimentation up to about 100 years ago, consistent with the catchment being closed tropical forest with little mineral runoff. This was transformed into rapid sedimentation rates this century as a result of forest clearing and small-scale but intensive agriculture.

In some examples, changes in the remains of organisms include the appearance and disappearance of species as well as just changes in abundance. In Lake Biwa, the natural oligotrophic phase of the past 75,000 years, which was characterised by two endemic species of diatom in the phytoplankton, *Stephanodiscus carconensis* and *Melosira solida*, began to break down in the early 1960s and a new diatom, the cosmopolitan species *Melosira italica* began to dominate the plankton with several other new species (Negoro, 1981). Changes in the centric diatom flora in the English Lake District over the last 20 years not only enable the patterns of present-day enrichment processes to be compared with earlier ones but also provide sediment 'markers' against which to measure future events (Haworth, 1979, 1985).

These examples and many other similar studies serve to provide a comparison of past enrichment events with present ones. They are important because of the extent of eutrophication and its emergence as an economic and political problem as well as a scientific one. They counteract the assumption that modern problems of eutrophication are comparable with earlier natural or semi-natural events which can be ignored because of their transience. In each of the above-mentioned lakes, limnological events linked to human activities, particularly in the past few decades, have been more extreme than any natural events recorded in the lakes' cores. Comparison between natural and artificial eutrophication can best be summed up in the words of Hutchinson (1973): 'Eutrophy in a New England lake before 1650 probably meant good fishing. Now it means a thick scum of blue-green algae.'

1.5 EXTENT OF ARTIFICIAL EUTROPHICATION

Eutrophication started to become a more widely used term from the late 1940s as scientists realised that plant nutrients entering and accumulating in lakes as a result of the industrial activities of the 20th century were causing changes in a matter of decades which naturally would occur over centuries or longer. Hasler (1947) summarised available knowledge for thirty-seven lakes in Europe and North America which had shown symptoms of enrichment over the past 200 years.

Sawyer (1947) was one of the first to show that changes in a chain of lakes in Wisconsin, USA, were related to urban and agricultural runoff. She documented changes in the treatment of sewage and industrial wastes at the town of Madison. Formerly served by a mixture of smaller septic tank and soakaway systems, the city began to construct a sewer system for waste collection in 1884 and then a chemical sewage treatment plant, discharging into Lake Monona, in 1899. From 1902 all sewage received some form of biological treatment. By 1918 algal nuisances in Lake Monona –

scums and odours – had become serious enough for the City Council to commission an engineering study. Copper sulphate treatment of the lake (copper sulphate is toxic to many algal species) was started in the same year. The engineering report discounted the effects of sewage treatment, suggesting that agricultural drainage was far more important as a cause of the algal blooms. Nevertheless, in 1926 a new sewage treatment plant was constructed on the outflow river to Lake Monona, which carried the effluent into the next lake in the chain, Lake Waubesa. In the 1930s Lake Waubesa began to develop algal nuisances and copper sulphate was introduced here. Continuing public discontent coupled with political disagreement over whether agricultural drainage or urban effluent was the major cause of the lake's deterioration led to a detailed study of nutrient sources between 1942 and 1944. This was the first nutrient budget study of a lake, and it revealed that at least 75% of the inorganic nitrogen and 88% of the inorganic phosphorus came from the sewage effluent inflow to Lake Waubesa.

Some of the largest lakes in Europe and North America have been well studied since the turn of this century and the sequence of changes provide clear demonstrations of the effects of artificial enrichment. In the Laurentian Great Lakes of America, data sets running for about a hundred years are available from analyses of water supply intakes and fish catch records, providing indirect evidence for eutrophication (Beeton, 1965, 1969). The chemical content of all lake waters with the exception of Lake Superior (which is the most northerly lake with the lowest human population in its catchment area) has changed, most markedly in Lakes Erie and Ontario, with increases in total dissolved solids, sodium, potassium, chloride, calcium and sulphate. The rate of increases in Lake Erie, the shallowest lake with the most densely settled catchment, have parallelled the rate of population increase. There have also been changes in the commercial fish populations (which can be used as an indication of eutrophication because Lake Erie is the only Great Lake whose fish populations have not suffered from the effects of the introduced sea lamprey *via* the Welland Canal). Catches during the 1910–1970 period of greatest change remained constant at around 23 million kg/ann, but species composition shifted markedly. Lake trout, whitefish, blue pike, sauger and cisco declined and were replaced by yellow perch, smelt, shepshead, whitebass and carp. A range of factors has contributed to these changes, including disappearence of shallow littoral spawning grounds by development, but the main cause has been the progressive development of oxygen depletion in the hypolimnion and consequent loss of benthic invertebrate species.

Lake Zurich, in Europe, has shown similar changes since the turn of the century (Hasler, 1947; Thomas, 1969). These have been particularly marked in one part of the lake – the Untersee – receiving the effluent from around 110,000 people. In the fish community there was a change in dominance from trout, coregonids and pike to cyprinid species even though the total harvest remained about the same. The effect of effluent was demonstrated by an increase in chloride concentrations from 1.3 mg/l in 1888 to 4.2 mg/l in 1916. Summer hypolimnion oxygen concentration decreased from 100% saturation at 100 metres depth between 1910 and 1930 to an average of 50% (and minimum of 9%) between 1930 and 1942. Algal changes occurred over this period, with increase in the bloom formation of species of *Oscillatoria* (cyanobacteria). Mean phosphorus concentrations have shown an increase from around 20 µg/l in 1946 to just under 80 two decades later, accompanied by further increases in the biomass of algae and nuisances caused by filamentous algal masses washed up on shorelines.

In every temperate country where urbanised areas are situated close to lakes, the period since the 1940s has seen the most rapid increase in detrimental changes taking place in these lakes. It is ironic that some of these lakes, such as those in Wisconsin quoted above, only began to show serious changes after municipalities began to organise sewage treatment works whose main effects (by efficiently breaking down organic wastes into soluble inorganic forms) were to provide lake algae with a ready assimilable supply of nutrients.

The problems associated with eutrophication are generally a more recent experience in tropical countries. It should be remembered however, that the benefits of deliberate eutrophication in the form of increased yields of edible fish from ponds fertilised with human wastes (Payne, 1984) have been enjoyed for centuries in Asia (Huet, 1970). The construction of new reservoirs, and the increased uses of natural lakes for water supply combined with settlement in their catchments has resulted in widespread problems following accelerated nutrient inputs. In Lake Valencia, South America, detrimental changes have occurred over this century following artificial lowering of lake levels combined with increases of total dissolved solids and nutrients following deforestation of the catchment (Serruya and Pollinger, 1983). Many lakes and reservoirs show increased eutrophication in southern Africa (Twinch, 1986) and in South America, where in countries such as Brazil the density of new reservoirs is very high (Tundisi, 1981). Some new reservoirs have begun life in a nutrient rich, deoxygenated state as a consequence of the decay of organic matter from their flooded basins (Mouchet, 1984). Nutrient runoff and sedimentation as a result of intensification of agricultural land use and settlement are problems likely to become widespread in many countries in the foreseeable future (Scheimer, 1983), but most of the present-day problems are caused by point sources from urban/industrial areas (Toerien, 1975). This makes them amenable to management and restoration using the same predictive models and technical restoration methods as in the temperate zone (Thornton, 1987) albeit with caution (Grobler, 1985; Cullen, 1986). The reasons for tropical and sub-tropical lakes behaving differently from temperate lakes in response to nutrients is usually associated with greater turbidities caused by more erratic rainfall and soils more prone to erosion, such that predicted plant growth is not realised. In New Zealand however, northern hemisphere models are not appropriate because nitrogen limitation occurs more widely than phosphorus limitation in lakes, due largely to greater application of phosphate fertilisers to pastures (White, 1983).

1.6 EUTROPHICATION IN RIVERS, ESTUARIES AND COASTAL WATERS

Lakes are points in the hydrological cycle where a mass of water is held up for a finite period. A number of changed environmental factors — chiefly reduced flow and turbulence coupled with increased light penetration — then combine to facilitate growth and production of plants. The nutrients are inevitably brought in by rivers and streams which themselves may show symptoms of eutrophication. Artificial eutrophication of rivers is a more widespread phenomenon than that of lakes; rivers worldwide have doubled their content of nitrogen and phosphorus as a result of the effects of man, with local increases in Western Europe and North America of up to 50 times (Meybeck, 1982). However, the effects of increased nutrient concentration on rivers has received far less attention. This is partly because lakes are more of a focus for human industrial and economic activity than rivers and partly because the biological response to elevated nutrient levels in rivers are less dramatic. For example, few rivers run slowly enough to produce a true phytoplankton and those that do tend to be dominated by small single-celled species which turn the water green rather than concentrate in floating mats or end up accumulated on the shoreline. Rivers are also far more likely to suffer from pollution – that is to say inputs of excessive quantities of organic matter whose metabolism grossly unbalances the river ecosystem. By the time they have recovered from this, either through the natural self-purification processes occurring with flow downstream, or through expenditure on improved effluent control, the effects of eutrophication are relatively unimportant unless the rivers flow into a lake or are of economic importance for recreation and fisheries, such as the rivers of Norfolk Broadland in England (Phillips, 1984).

Nevertheless, the effects of eutrophication are often apparent to the layman and scientist alike as increased growths of tolerant macrophyte species such as *Potamogeton pectinatus* or of filamentous algae (Lund, 1970; Bolas and Lund, 1974) which may themselves cause night-time deoxygenation and have a consequent effect upon animal communities. Animal species may change in response to the changed proportions of food available as allochthanous detritus is replaced by more autochthanous algae, bacteria and fungi. Fish species may change if oxygen-sensitive species are affected by the night-time oxygen sag. Overall, enrichment tends to produce biological effects naturally found in the lower reaches of rivers, higher upstream (Hynes, 1969).

At the downstream end of rivers and lakes are estuaries. These are naturally the most productive ecosystems in the world because they are shallow and well-mixed, receiving nutrients from rivers, coastal waters and adjacent marshes and swamps (Goldman and Horne 1983). The problems of over-enrichment of estuaries were addressed later than those of lakes, with initial work attempting to utilise the experiences from lake studies (McErlean and Reed, 1981) in dealing with estuarine problems. There are many differences in estuaries however, notably the two-way transport and current processes as a consequence of the interplay of freshwater flow and tidal surge. These control the extent of salinity, resuspension of sediment, water column mixing and oxygen regimes. Species of phytoplankton are different and show different nutrient responses; for example the cyanobacteria, bloom-forming algae dominant in eutrophic waters of low salinity are replaced by dinoflagellate, 'red-tide'-forming species in higher salinity (Ryther, 1981). These may cause unpredicable conditions of anoxia in areas where they accumulate leading to mortalities of fish and invertebrates (Chiaudani *et al.*, 1986).

1.7 MEASUREMENT OF EUTROPHICATION

Throughout the history of eutrophication studies, from the early classification of lakes to the development of predictive and management models, many of the manifestations of eutrophication have been expressed qualitatively. Descriptions such as the transparency of the water, or the occurrence and frequency of 'nuisance blooms' are still valid as an expression of the impact of eutrophication to the non-scientist, but development of the scientific understanding of eutrophication can be seen as a progression from qualitative measures of the causes and their effects to quantitative ones. This progression is implicit in the explanation of the phenomenon of eutrophication in the remainder of this book.

- 2

The nutrients causing eutrophication, and their sources

2.1 THE REQUIREMENTS OF LIVING CELLS FOR SURVIVAL AND GROWTH

The previous chapter referred rather generally to nutrient increases as the cause of both natural and accelerated eutrophication in order to introduce the concept and its effects. We can now discuss the evidence for the relative importance of individual nutrient elements in this process.

Living organisms require around 40 of the elements which naturally occur in the earth's crust and atmosphere to sustain growth and reproduction. The most important, carbon, is usually considered separately from the others, because it is the energy locked into chemical bonds between carbon atoms and those with oxygen and hydrogen atoms which is the basis of the photosynthetic conversion of solar energy into living tissue. Oxygen and hydrogen are freely available in water under most circumstances. Other essential elements are usually considered in two groups: the macronutrients or major elements, required in large quantities, and the micronutrients or trace elements, required in small quantities. Calcium, magnesium, potassium, nitrogen, phosphorus, sulphur and iron are the most important of the macronutrients, together with silicon (used in cell frustules by diatoms and a few other algal species), whilst copper, cobalt, molybdenum, manganese, zinc, boron, vanadium, chlorine and vitamin complexes are the most important of the micronutrients.

Of all the elements derived from the earth's crust (the 'lithosphere') present in plant tissue, phosphorus and selenium are those whose proportional abundance is lower in the lithosphere than in plant tissue (Hutchinson, 1973). Phosphorus is thus a prime candidate for a limiting macronutrient. Selenium, followed by zinc, molybdenum and manganese are potentially likely to be limiting micronutrients. Of the elements derived from the atmosphere or hydrosphere – carbon, nitrogen, oxygen and hydrogen – nitrogen might be limiting because its major reservoir is the gaseous form unusable by plants directly whereas carbon is unlikely to be limiting because its main reservoir − dissolved carbon dioxide gas − is soluble in water.

2.2 THE IMPORTANT LIMITING NUTRIENTS

Initial studies on the effects of artificial eutrophication did not distinguish between possible limiting factors but cited increases in several nutrients as the cause of observed changes in plant productivity. Hasler (1947) referred to all nutritive substances, especially N and P, whilst Sawyer (1947) compared the effects of sewage enrichment of lakes with effects deliberately created by the addition of commercial N-P-K fertilisers on crops, or of soy bean and cottonseed meal to lakes. She subsequently showed that phosphorus was probably the key element controlling productivity in the Wisconsin chain of lakes, with nuisance algal blooms occurring when average inorganic phosphorus concentrations were higher than 10 μg/l. Nitrogen also played a role, with critical levels around 300 μg/l inorganic nitrogen. Sawyer found after careful study of the composition of urban and industrial drainage inflows that biologically treated sewage contained 200 times these nutrient concentrations and agricultural drainage about twice as much. However, the relative proportions of the two nutrients in the drainage inflows were different. Agricultural drainage contained 18 times the proportion of nitrogen to phosphorus, treated sewage effluent only about 5 times as much. Thus in both sources, but particularly the latter, phosphorus was supplied in quantities and ratios greater than the minimum needed to stimulate algal growth.

In his own work on Swiss lakes, and from comparisons with other studies, Thomas (1969) demonstrated that phosphorus was the major element whose increase in concentration promoted undesirable growths of algae in many European lakes during the first half of this century. In oligotrophic lakes phosphorus concentrations were very low but during the summer nitrate–nitrogen levels were high because phytoplankton could not utilise the available nitrogen under conditions of phosphate limitation. Subsequent increases in phosphorus concentrations caused summer depletion of nitrogen by enhanced algal growth. Laboratory experiments with additions of phosphate and nitrate confirmed these conclusions, showing that addition of phosphate and nitrate without other nutrients was enough to increase phytoplankton in Swiss lakes.

Laboratory studies with single species of bloom-forming algae indicated which nutrients were limiting their growth. Gerloff (1969) showed that addition of phosphate, nitrate and iron produced enhanced growth of *Microcystis aeruginosa* in water from Lake Mendota which was slightly more than that produced by any other combination of elements including all the essential elements together. Comparison with growth in culture vessels where the nutrients were added individually indicated that in this experiment, nitrogen was the initial limiting nutrient, followed by phosphorus and then iron.

Nitrogen has been found to be limiting in many African lakes for most of the year (Viner, 1975) because nutrients are low in well-leached tropical soils, efficiently locked-up by savannah grass communities, and those which do leach are absorbed by fringing swamp communities. Talling found low levels of nitrogen compared to phosphorus in East African freshwater lakes with probable seasonal nitrate limitation of phytoplankton in Lake Victoria associated with the cycle of stratification (Talling, 1966; Talling and Talling, 1965). Viner showed permanent shortage of nitrogen and phosphorus in Lake George, Uganda, with nitrogen supply being more critical (Horne and Viner, 1971; Viner, 1977). Moss showed nitrate, phosphate and sulphate to be limiting algal growth in Lake Malawi, in that order (Moss, 1979a). In South America, Lobo reservoir, Brazil, showed primarily nitrogen-limited phytoplankton growth, with phosphorus enrichment of *in situ* experiments producing no growth (Henry *et al.*, 1984).

In some temperate lakes, nitrogen may become temporarily limiting in summer but this often causes a shift towards colonial species of cyanobacteria in the plankton which are directly capable of fixing atmospheric nitrogen in special cells called 'heterocysts' (Stewart, 1969). In lake fertilisation experiments, Schindler (1977) showed that reducing the ratio of nitrogen to phosphorus in fertilisers added to two separate lakes resulted in the appearance of nitrogen-fixing cyanobacteria of the genera *Anabaena* and *Aphanizomenon*. In the majority of temperate lakes, phosphorus is the primary limiting nutrient and its availability controls the use of nitrogen and carbon from the atmosphere. Silicon is limiting for diatom growth under certain conditions; Lund showed for example that for *Asterionella formosa*, which is frequently a nuisance diatom to British potable water treatment works, population growth did not occur in Lake Windermere, English Lake District, at silica levels below 0.5 mg/l as Si (Lund, 1950). Laboratory culture showed that addition of small amounts of phosphate resulted in almost complete silica depletion by the subsequent diatom growth, indicating that phosphorus in fact controlled silica uptake (Hughes and Lund, 1962). Lund subsequently showed that increased input of phosphorus to Blelham Tarn between 1945 and 1967, which was not accompanied by increases in nitrogen or silicon, was followed by increased diatom populations and decreased silica concentrations (Lund, 1969). Work on Lake Michigan waters has shown that progressive increase in phosphorus concentrations during the course of this century has resulted in greater depletion of the silica pool through enhanced diatom growth (Schelske and Stoermer, 1972).

Carbon is the major element in living tissue, and it is present in algal cells in proportion to nitrogen and phosphorus in a ratio of about 106:16:1 (Redfield, 1934). Diffusion of carbon dioxide from the atmosphere is the major source and the large reservoir in the atmosphere, coupled with its ready solubility in water, means that it is rarely limiting for algal growth. For a period in the late 1960s however, it was suggested that diffusion from the atmosphere was inadequate and that the production of carbon dioxide by bacteria regulated algal productivity (Kuenzel, 1969). Some experiments with the addition of carbon, nitrogen and phosphorus to enclosures resulted in rapid bacterial growth enhancing dissolved carbon dioxide and bicarbonate concentrations (Kerr *et al.*, 1970, 1972). The authors suggested that it was the bacterial production of carbon, rather than inorganic nutrients, which limited algal growth. These experiments were criticised, however, on the ground that they contained excess phosphorus (1–2 mg/l) and low concentrations of carbon dioxide without suitable controls. Moreover, the conclusions were not widely repeated when similar experiments were carried out on a range of lake water types; algal growth rates were produced which were directly proportional to the amount of phosphorus added but not to the amounts of carbon or nitrogen (Maloney *et al.*, 1972). From experiments like these and from studies on photosynthesis of natural populations of algae (Talling, 1976) it was generally concluded that carbon limitation of algal production only occurs at the height of photosynthesis in low alkalinity waters and that limitation only becomes important when these naturally infertile waters experience enhanced levels of nitrogen and phosphorus. At the other end of the scale of enrichment, highly fertilised sewage treatment lagoons produce such high biomasses of algae that the rate of supply of carbon dioxide may become limiting before nitrogen and phosphorus are exhausted (King, 1972). These conclusions were reinforced in a wide ranging review of the role of carbon sources on algal growth in relation to eutrophication effects (Goldman *et al.*, 1972).

Final resolution of the controversy over the importance of carbon limitation of algal growth came primarily from a series of enrichment experiments of whole-lakes in the nutrient-poor Pre-cambrian Shield lake District of Canada. In an initial experiment, phosphate, nitrate and sodium were added over a 17-week period in summer 1969 to a small lake, 'Lake 227' (Schindler *et al.*, 1971). Phosphate was rapidly taken up by the living particulate matter (seston) – bacteria and algae. Levels of dissolved phosphorus did not exceed 8 µg/l compared to pre-fertilisation levels of up to 5 µg/l whereas particulate concentrations reached 29 µg/l compared with 5 µg/l. Added nitrate was also assimilated by seston, with a maximum of 600 µg/l compared with pre-fertilisation levels of up to 100 µg/l. Dissolved nitrate, in contrast to phosphate, also increased in concentration. The particulate and dissolved forms of both phosphorus and nitrogen increased in concentration in the hypolimnion during the period of stratification in July and August, as might be expected if they were first incorporated into the growth of living cells and then subsequently sedimented and released after death and decay of the individual organisms. In contrast, sodium accumulated in the epilimnion and did not move into the hypolimnion until after overturn, suggesting that it was not utilised in growth. Other major ions did not show concentration changes any different from comparable unfertilised lakes.

Phytoplankton standing crop increased after fertilisation to a maximum of 52 µg/l measured as chlorophyll 'a' compared with 3 in the year before and 9 for the highest in a range of unfertilised adjacent lakes. Species succession of algae changed, with over-whelming dominance of green algae (chlorophyta) and an increase in blue-greens (cyanobacteria) after fertilisation. Species in these two groups were never formerly abundant; phytoplankton of these lakes was normally dominated by species of brown algae in the groups cryptophyceae, chrysophyceae and bacillariophyceae (diatoms) (Fig. 2.1).

Additional experiments were carried out in enclosures to assess whether carbon was limiting phytoplankton growth. A first set of four polythene tubes, with carbon added in various forms (gaseous, inorganic and organic), was established in the lake in early August as carbon dioxide concentrations in the lake dropped below 20 μmoles/l. In all tubes, phytoplankton growth was lower than in the lake. In a second set, combinations of nitrogen, phosphorus and carbon were added. The tube receiving only carbon and nitrogen (as urea) produced similar chlorophyll concentrations and phytoplankton volumes as the control. All the other tubes produced increased algal crops; the tube receiving phosphorus alone showed a less sustained increase but all other combinations of nutrients produced about twice the chlorophyll concentrations and phytoplankton volumes as the control. In the final two tubes, human urine was added and this produced around twice the chlorophyll biomass of the other tube additions – over 300 μg/l. The further addition of bicarbonate to one of the urine-enriched tubes produced a standing crop in excess of 350 μg/l. These experiments showed that the addition of phosphorus and nitrogen alone to a lake would cause its phytoplankton biomass to increase, and that carbon was not limiting the increase even when

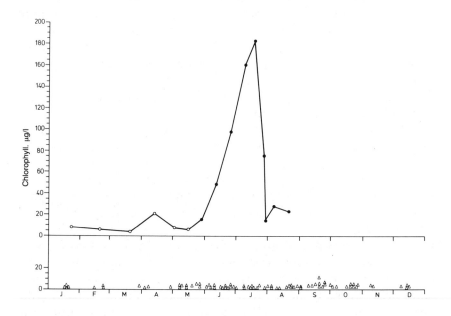

Figure 2.1 A comparison of algal growth measured as chlorophyll 'a' in Lake 227 in 1973 after the addition of phosphorus and nitrogen (addition after open circles change to closed) compared with a range of unfertilised lakes. Modified from Schindler and Fee (1974), with permission.

levels of carbon dioxide were reduced to 20 μmoles/l. Only at the very high levels of chlorophyll in the last enclosure was there indication that carbon limited the biomass achieved.

Three further years of enrichment in the lake (Schindler *et al.*, 1973) produced phytoplankton standing crops close to those considered to be the theoretical maximum which may be sustained by surface light penetration. Particulate carbon levels increased each summer concomitant with the increases in particulate phosphorus and nitrogen, even though no carbon was included in the fertiliser additions. Carbon budget calculations showed that between 70 and 95% of incoming carbon was derived from the atmosphere, with the lake 'soaking up' carbon dioxide as needed to sustain the phytoplankton biomass which was generated by the phosphorus and nitrogen additions.

The role of carbon was examined in more detail during a short summer period when the lake's inflow streams were dry (Schindler *et al.*, 1972). Phosphorus and nitrogen were added, but no carbon, and budgets calculated for the beginning and end of the 14-day period. To account for the added phosphorus taken up by the seston and sedimented, an estimated 0.2 g of carbon per m^2 per day was needed. This could have been adequately supplied by diffusion across a concentration gradient from the atmosphere to the lake; calculation of partial pressures of carbon dioxide in the lake water from pH measures showed that such a gradient existed during July and August. The rate of carbon dioxide movement was also shown to be possible in experiments which measured the diffusion of inert radon gas from water to atmosphere.

Confirmation of the unimportance of carbon as a limiting factor in the absence of phosphorus was then obtained by experiments with another lake, 'Lake 226', in 1973. This contained two basins separated by a narrow channel which was artificially blocked off with a thick vinyl curtain (Schindler, 1974). Both basins were fertilised by adding nitrate and sucrose, but one was also given phosphate. The basin with nitrate and sucrose alone showed no significant increase in algae, whereas that with additional phosphate developed dense blooms of algae. The ratio of nitrogen to phosphorus used in Lake 226 was lower than in Lake 227, close to 5:1 compared with 14:1 in the latter. The result was dominance of the phytoplankton by a nitrogen-fixing species which restored the nitrogen:phosphorus ratio to one similar to other lakes (Schindler, 1977).

Schindler considered that any limitation of phytoplankton growth by micronutrients was an unlikely occurrence; he likened lakes generally to 'very dirty glassware' (Schindler, 1981) when referring to the careful culture studies necessary to detect trace element limitation. Only in lakes where catchment soils were deficient in one or more trace elements has limitation of algal growth been clearly demonstrated, such as Castle Lake, California, deficient in molybdenum (Goldman, 1960) and some lakes in New Zealand deficient in cobalt and zinc (Goldman, 1964).

2.3 THE SUPPLY OF NITROGEN AND PHOSPHORUS TO LAKES

2.3.1 Introduction

In a natural, undisturbed environment the nutrient supply of a lake will be derived from the drainage of the catchment together with direct rainfall on the lake surface and any internal recycling which may occur from the sediments. Studies which have been made upon such catchments (which in the northern hemisphere are almost entirely forested) have shown that nutrient runoff is very low (Borman and Likens, 1967; Bormann *et al.*, 1968; Hobbie and Likens, 1973) because cycling within the vegetation of the terrestrial ecosystem is very tight. The same is true of tropical forests and savannahs (Nye and Greenland, 1960; Viner, 1975). In the temperate zones, runoff from natural or secondary grassland is higher in nutrients than runoff from forested land, and runoff from arable land is higher still. Urban areas and effluents produce a range of high-nutrient effluents.

2.3.2 Natural background sources in catchments

The initial natural source of phosphorus is weathering of phosphate-containing rocks. Igneous rocks contain apatite – complexes of phosphate with calcium – the weathering and subsequent marine sedimentation of which has given rise through geological history to phosphates widely distributed in sedimentary rocks. The common weathering processes of such rocks lead to clays in which the phosphate is moved from apatite into the clay complex. It is both tightly bound into the clay lattice in place of hydroxyl ions and more reversibly bound by electrostatic attraction to aluminium or iron ions (Golterman, 1975; Holtan *et al.*, 1988) (Fig. 2.2).

Nitrogen too is present in igneous rocks at low concentrations but the main source of nitrogen for all biological activity on this

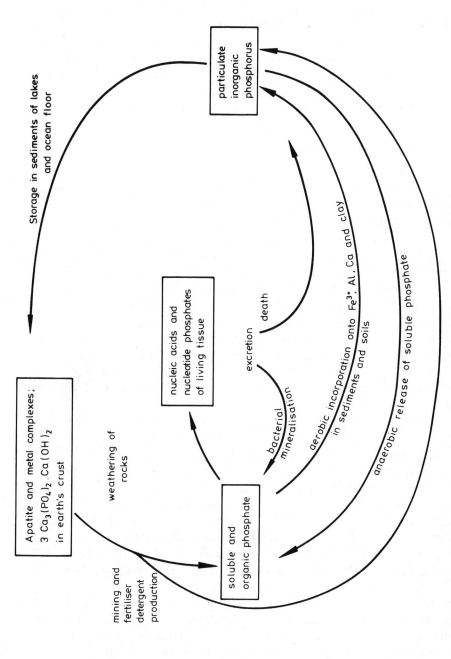

Figure 2.2 The major pathways and transformations of phosphorus in the environment.

planet is the atmospheric reservoir of gaseous nitrogen. Nitrogen gas is chemically very stable but is made available to organisms by fixation into a variety of oxides or reduction to ammonium. These events do occur as a result of electrical or photochemical processes in the atmosphere but the major pathway is fixation by microorganisms in the soil, which is about seven times greater than nitrogen from all atmospheric processes brought to earth by rainfall (Hutchinson, 1944; Larcher, 1975) (Fig. 2.3).

2.3.3 Atmospheric sources

Atmospheric input may be an important source of both nutrients, particularly in dilute lakes in remote, igneous rock areas such as northern Canada (Barica and Armstrong, 1971). Phosphorus in rainfall is particulate material washed out of the atmosphere which may be derived from sources (pollen, dust, soil particles etc.) far away from the receiving catchment (Fig. 2.4). Levels of total (wet plus dry) atmospheric phosphorus deposition were estimated to be 270 g P/ha/ann in Sweden (Odén and Ahl, 1976) but higher in the United States (Rigler, 1974). A recent summary of 16 studies in North America and Europe gave an average of 430 g total P /ha/ann (Holtan *et al.*, 1988), with the lowest levels (of 50) probably reflecting natural background levels (Ahl, 1988). Estimates of direct weathering and runoff of phosphorus from underlying rocks are difficult to obtain before the phosphorus passes through the vegetation and soils of catchments.

Nitrogen deposition, which includes the products of chemical fixation as well as particulate sources, is at least an order of magnitude higher, between 2 and 10 kg total N/ha/ann (Brezonik, 1972), but may be up to 30 kg when global terrestrial dry deposition of ammonium and oxides of nitrogen are added to wet deposition estimates (Söderlund and Svensson, 1976). Atmospheric input of nitrogen has been far more extensively affected than that of phosphorus by human activities because many of these put gaseous nitrogen forms into the long-range atmospheric circulation patterns (Fig. 2.5). Chief among these are gaseous ammonia from the decomposition of human and animal wastes, and gaseous oxides of nitrogen from the combustion of fossil fuels (Söderlund and Svensson, 1976). In the United Kingdom, rainfall input of total nitrogen at Rothamstead, in south-east England, has increased from a maximum of 5 kg N/ha/ann up to 1913 (Russell and Richards, 1919) to a maximum of 18.2 kg N/ha/ann in the decade 1969–78 (Anon., 1983).

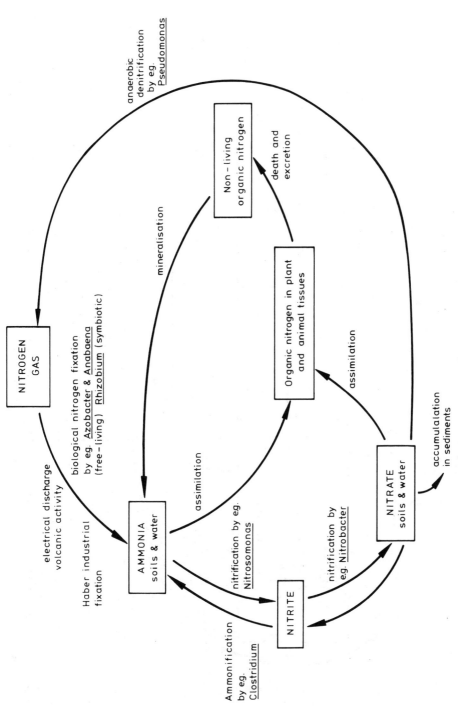

Figure 2.3 The major pathways and transformations of nitrogen in the environment.

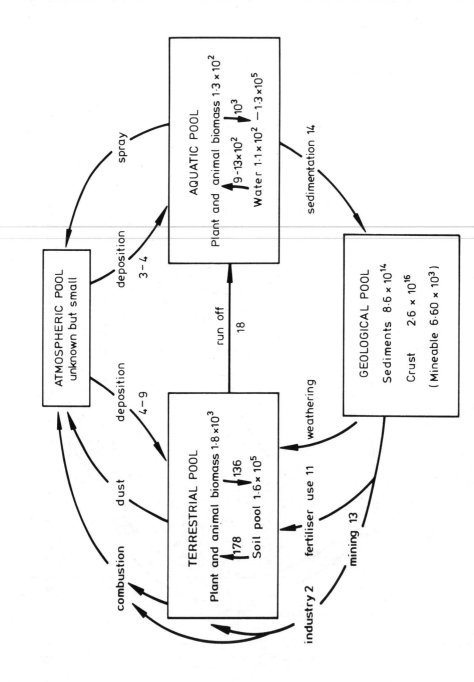

Figure 2.4 The global phosphorus cycle including both natural and human-enhanced pathways. Quantities taken from Pierrou (1976), reservoirs in Tg (10^{12} g), fluxes in Tg/ann.

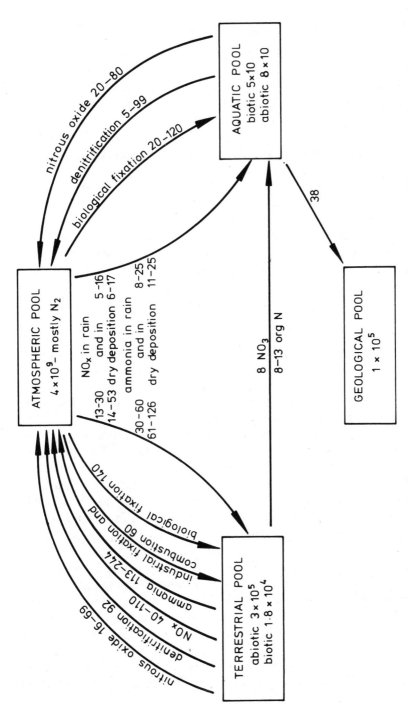

Figure 2.5 The global nitrogen cycle including both natural and human-enhanced effects. Quantities taken from Söderlund and Svensson (1976), reservoirs in Tg (10^{12} g), fluxes in Tg/ann.

2.3.4 Nutrient storage in soils and losses by leaching

Both elements are found in soils predominantly as organic compounds in either detritus (humus) or living tissue. The quantities are very variable, depending upon such factors as the vegetation cover, the quantity of detritus on or in the soil and microbial activity in the soil releasing inorganic compounds.

The chemical behaviour of the inorganic forms of the two elements is quite different and this has consequences for losses from the soils. Phosphate ions are electrostatically bound to the molecular lattice of soil particles (Golterman, 1975), particularly those containing more than 5% clay (Cooke and Williams, 1973) and are thus not easily leached out of most soils in soluble forms, even those heavily fertilised. However, drainage from eroded soils carries higher amounts of phosphorus than from vegetated soils because the phosphorus remains bound to soil particles. Surface runoff contains higher concentrations of total phosphorus than does infiltration water (Williams, 1971) because of its higher content of suspended particles.

Nitrate or ammonium ions by contrast, are not tightly bound. Ammonium ions are adsorbed onto clay minerals and nitrate is in solution in soil moisture; both are thus easily leached out of the upper soil layers by rainfall. Nitrogen compounds are more readily lost from lighter, well-drained soils than from heavy soils and more from arable land than from grassland.

Comparison of the nutrient concentrations in the inputs and outputs of different experimental catchments has shown the extent of natural fluxes. Undisturbed forested catchments in the Hubbard Brook ecosystem, North America, lost very little phosphorus in streamflow, 21 g P/ha/ann; less than was added annually by rainfall, 108 g P/ha/ann (Hobbie and Likens, 1973). Both input and output are small compared with the quantity cycled within the living component of the terrestrial ecosystem, with annual leaf fall averaging 1.9 kg P/ha/ann. Nitrogen input and output were both greater, with a seasonal pattern of high levels in the winter months, low in summer. Input was about 5 kg N/ha/ann and streamflow 2 kg N/ha/ann and both were again only a small fraction of the annual cycling within the system – 3.2×10^3 kg N/ha/ann (Bormann *et al.*, 1968).

These nutrient losses are similar to those found in forested catchments in Scandinavia. An output of 30–90 g P/ha/ann together with 0.9–1.6 kg N/ha/ann was recorded from a seven year study in

Sweden (Ahl, 1975). Undisturbed tropical catchments, in Amazonia (Uhl and Jordan, 1984) and the forested Ruwenzori mountains in Africa (Viner, 1975) have similarly low nutrient outputs which match or are less than inputs.

Differences in geology may influence the extent of nutrient losses. Dillon and Kirchner (1974) analysed phosphorus export from 31 different lake catchments in southern Ontario, Canada, and reviewed the literature of phosphorus losses from north temperate catchments for comparison with their own. They were able to show significant differences in phosphorus loss between forested igneous catchments (mean loss 48 g P/ha/ann) and forested sedimentary rock catchments (mean loss 107 g P/ha/ann). Losses from forested volcanic catchments were higher still, at 720 g P/ha/ann. They were also able to show differences between intact forest catchments, of either rock type, and forests with pasture comprising more than 15% of their catchments. In Ontario forests, this slight increase in intensification of land use increased phosphorus export to 117 g P/ha/ann on igneous catchments, 288 g P/ha/ann on sedimentary.

2.3.5 The effects of land use changes on nutrient losses from diffuse sources

Intensification of land use by human activity inevitably disturbs the naturally conservative cycling of nutrients in the terrestrial eco-system and increases their concentration in runoff. Losses increase in proportion both to the intensity and the frequency of disturbance.

In the Hubbard Brook ecosystem, clear-cutting of a forest block and inhibition of vegetation re-growth by herbicide at the beginning of the annual growing season resulted in large losses of nitrogen in the runoff streams, up to a maximum of 60 mg/l compared to <1 mg/l from an adjacent undisturbed catchment (Bormann *et al.*, 1968) (Fig. 2.6). Annual losses of 53 kg N/ha compared with a net gain of 3 kg/ha in undisturbed forest. Phosphorus runoff was less dramatic, about twice as much soluble and fine particulate phosphorus was lost compared with the undisturbed catchment, but twelve times more in larger (>1 mm) organic and inorganic particles due to higher erosion in the deforested catchment. Overall the annual loss of phosphorus was estimated to be 107 g P/ha compared with a gain of 87 g P/ha in undisturbed forest.

Clearance of tropical forest resulted in similar rapid increases in the concentration of nitrate–nitrogen in runoff (Uhl and Jordan, 1984). In both experimental disturbances, nitrogen concentration in

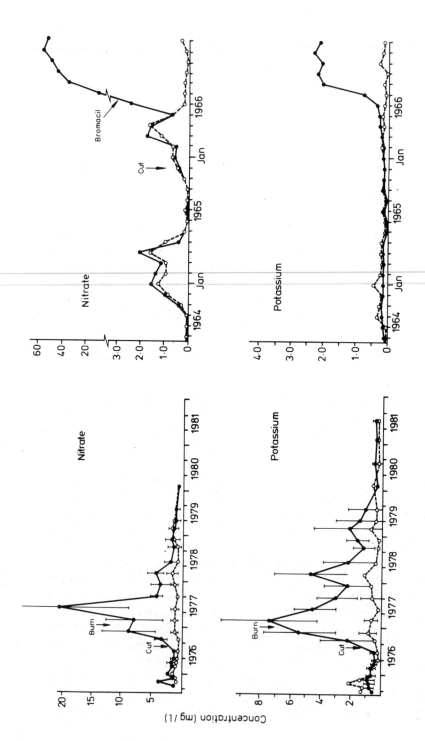

Figure 2.6 Losses of nitrate and potassium from experimental clear-felling and subsequent burning or herbicide application from a tropical (left) and temperate (right) forest. Modified from Uhl and Jordan (1984) and Bormann *et al.* (1968), with permission.

runoff declined to the low levels of undisturbed forest within three years after vegetation regrowth was allowed to occur (Bormann *et al.*, 1974; Uhl and Jordan, 1984). This was due to a combination of the uptake of nutrients into new plant tissue coupled with a reduction in the amount decayed and leached from the original cut vegetation.

Most of man's agricultural use of land which was once naturally forested differs from these experimental disturbances in one important aspect – the changes are more or less permanent and the disturbance usually regular. The comparisons of phosphorus runoff by Dillon and Kirchner indicate the consequences of this. Pasture areas represent the first step in a range of human disturbances, in this case the replacement of natural forest vegetation cover by another permanent or semi-permanent vegetation cover based upon native grasses and herb species, maintained by the activities of domestic grazing animals. Nutrient runoff from permanent pastures is higher than forested catchments, about twice as much phosphorus and nitrogen depending upon the type of soils. A range of nitrogen losses from 1 to 9 kg N/ha/ann have been recorded from low-intensity grasslands (Ahl, 1975; Gachter and Furrer, 1972). More intensively used pastures, particularly in lowland regions, are frequently fertilised to increase plant growth. Studies of fertilised grasslands in England have indicated losses equivalent to 2–5% of added nitrogen fertiliser but in areas of high rainfall (Lancashire, 1500 mm/ann) leaching losses reached 250 kg N/ha/ann, equivalent to 10–40%. At high rates of fertiliser addition, losses may also increase to 142 kg N/ha/ann for an application of 500 kg N/ha/ann (Anon., 1983). The soil type influences losses from grasslands as from forests; average annual losses from unfertilised grasslands on Silurian parent material were 6.2 kg/ha, three times the loss from igneous rock catchments (Anon., 1983).

Further intensification of agricultural land use results in greater nutrient losses. The precise quantities lost depend on various factors; chief amongst which are soil type, soil drainage, rainfall, nature and extent of vegetation cover, nature and amount of applied fertilisers and density of domestic animals. The latter four factors also change with season and from year to year. Nitrogen losses increase in proportion to intensity of use and to artificial additions more so than phosphorus losses, because of the greater mobility of soluble nitrogen compounds. In lysimeter studies at Rothamstead, with fallow soil and no fertiliser additions, the equivalent of 23 kg N/ha/ann was lost; lysimeters in Aberdeen, north-east Scotland, with around 250 kg N/ha/ann added in fertiliser, lost up to 52 kg/ha/ann (Cooke and Williams, 1973). Some lysimeter studies have recorded average annual losses of up to 200 kg N/ha on unfertilised sandstone soils, rising to 300 after fertilisation (Anon., 1983). These can be regarded as upper extremes, since even with annual arable crops, the soil surface is covered by vegetation for more than half the year, so that plant root uptake reduces nutrient loss. In an intensively farmed arable catchment in north-east Scotland, whose crops included raspberry canes and rhubarb (with large areas of bare soil between rows), annual nitrogen losses were around 100 kg N/ha (Stewart *et al.*, 1975). In single arable crop studies, such as fertilised corn crop catchments in Missouri USA, annual losses ranged from 10 to 34% of added nitrogen (equivalent to 40–57 kg N/ha/ann) (Alberts *et al.*, 1978; Schuman *et al.*, 1973).

Phosphorus losses from intensively farmed crops remain comparatively low unless there is soil erosion or waterlogging. A mean loss calculated for arable soils in the Netherlands was 250 g P/ha/ann (Kohlenbrander, 1972) whilst lysimeter losses in the UK ranged from 70 to 250 g P/ha/ann (Cooke and Williams, 1973). In the Missouri catchment referred to above losses recorded were up to 1.9 kg P/ha (6% of added P) of which 1.8 kg P/ha was particulate, associated with sediment erosion, and only the remaining 100 g soluble phosphate.

Waterlogging makes phosphorus more soluble; soils in California under rice cultivation lost up to 530 g P/ha/ann compared with only 80–200 g under lucerne (Johnston *et al.*, 1965). Increase in slope causes relatively more phosphorus to be lost than nitrogen because it increases particulate runoff; a slope increase from 8° to 20° increased phosphorus losses by 360% (from 450 g P/ha) but only doubled nitrogen losses (from 16 kg N/ha) (Goldman and Horne, 1983). Erosion on steep uncultivated land or exposed arable land has been estimated to result in topsoil losses containing as much as 6–12 kg P/ha/ann, and in extreme cases of wind erosion, up to 150 Kg P/ha and ten times as much nitrogen (Cooke and Williams, 1970, 1973).

Forestry operations are increasingly becoming more commercial and intensive as world demand for wood products increases; extension of planting on up to 1.8 milion hectares have been suggested for the UK (Anon., 1980). Many new plantations are established on low-grade agricultural land and require fertilisation (Youngman, 1986). Studies of nutrient runoff from fertilised forest catchments have shown losses of total phosphorus averaging 2 kg P/ha/ann compared with 0.12 in control catchments and an average 14 kg N/ha/ann compared with about half this value in controls (Harriman, 1978).

Losses of both nitrogen and phosphorus are seasonal, depending upon rainfall patterns and the cycle of crop planting and harvesting. Alberts *et al.* (1978) showed that 51 kg out of the 57 kg N/ha and 1.3 kg P out of the 1.9 kg P/ha lost annually occurred during the April–June period of planting and fertilising. Studies in an Illinois corn belt catchment using relative enrichment of nitrogen-15 to identify the sources of N in tile drainage indicated that at least 55–60% of nitrate runoff during the spring period (April–July) originated as added fertiliser, compared to 20% over the rest of the year (Kohl *et al.*, 1971).

Studies which have focussed upon the nutrient losses from single crops are less common than studies of whole lake catchments under mixed agriculture. Nevertheless the pattern of mixed agricultural catchment losses revealed by measurements at the point of exit of the catchment (usually lake inflows or stream confluences) is broadly similar to losses revealed by single-crop studies. In a catchment of mixed pasture (around 70% and arable (30%) in Devon, south-west England), Troake *et al.* (1975) found annual nitrate losses of 25–30 kg N/ha (which represented 26–50% of added nitrogen fertiliser). A similar catchment in Eastern Scotland, Loch Leven, which is 71% grassland and 29% arable (Smith, 1974) lost 33 kg N/ha/ann (which represented around a third of applied fertiliser) and 200 g P/ha/ann (Holden, 1975).

In most industrialised countries, high nutrient losses from agricultural soils are a comparatively recent phenomenon. Up to about 50 years ago most fertilisation of pastures was done with farmyard manure. This method of fertilisation, even when applied to fallow surfaces, rarely results in greater nutrient losses than the unfertilised soil. Losses increase when fertilisation is applied as slurry (from concentrated livestock units) as well as artificial fertiliser, the two sources most commonly used today in modern farming operations (Anon., 1983). There has been a six-fold increase in the use of nitrogen fertilisers in the United Kingdom between 1930 and 1980 (Anon., 1983) (Fig. 2.7) and a similar increase in the United States (Porter, 1975). It is a reflection both of the changeover in fertiliser type (from manure to artificial) and the change in land use (from pasture to arable). Over the same period there has been a steady increase in concentration of nitrogen in

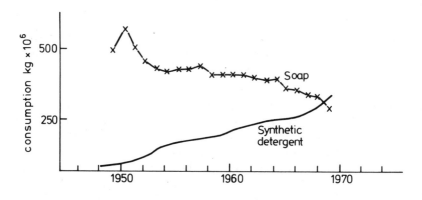

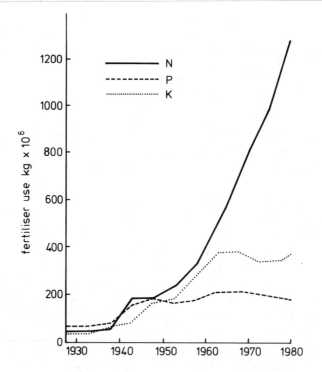

Figure 2.7 The increase in human use of detergents relative to soap (upper) and fertilisers (lower) in the UK in recent decades. Modified from Deevey and Harkness (1973) and Anon. (1983), with permission.

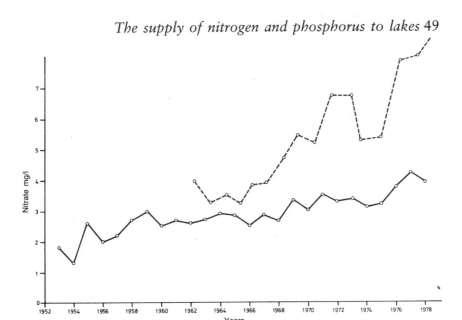

Figure 2.8 Recent increases of nitrate-nitrogen in the rivers Great Ouse (predominantly arable – dotted line) and Dove (predominantly pasture – continuous line) in lowland England. Modified from Wilkinson and Greene (1982) and Brierley (1985), with permission.

rivers draining agricultural land (Fig. 2.8).

The use of phosphorus in fertilisers, in contrast to nitrogen, has remained relatively static in the UK since the 1940s, primarily because it is the supply of nitrogen to growing plants at the right time which limits agricultural production (Anon., 1983). However, in some countries such as Ireland, phosphorus fertiliser application has increased steeply and increased the available phosphorus in soils but not appreciably increased runoff losses (Hanley and Murphy, 1973). In all countries with intensive agriculture, phosphorus losses from agriculture and inputs to aquatic ecosystems have continued to increase as a consequence of changes in animal husbandry. Both an increase in stocking density of free-ranging animals and an increase in total number of animals maintained in battery units have occurred particularly over the past 30 years (Porter *et al.*, 1975b). Animals maintained in battery units are now more usually fed silage (composted grass) rather than hay (dried grass) which results in a more liquid, slurry waste. In the UK about half of the total 106 metric tonnes of nitrogen estimated to have been excreted by livestock in 1978 was voided directly onto land, the other half was produced in concentrated units: this total animal production was approximately equal to annual fertiliser nitrogen use (Anon., 1983). Approximate UK production of phosphorus in animal excreta is 2 × 105 metric tonnes per year, again about equal to fertiliser use (Vollenweider, 1968).

There are two consequences for nutrient runoff as a result of these trends. The first is that increasing quantities of animal wastes may be applied to land in high quantities or in unwise fashion (Hanley and Murphy, 1973), leading to elevated concentrations of nitrogen and phosphorus in drainage water. The second is that intensive animal units become 'point sources' of nutrient runoff rather than 'diffuse sources'; they may then contribute problems of organic pollution as well as nutrient enrichment to the receiving watercourses. The largest growth of water pollution incidents in the United Kingdom over the past two decades has been from agricultural point-sources such as livestock or silage store units (Howells and Merriman, 1986). Such units often contain nutrient concentrations greatly in excess of human sewage and in some agricultural areas the total nutrient quantities far exceed those of humans; in South Dakota, USA, for example, the livestock population produced manure equivalent to 64 million humans whilst the actual human population was three-quarters of a million (Vollenweider, 1968). Altogether farm animals produce ten times as much waste in the USA as does the human population (Cooke, 1975). Spreading of animal wastes and manure to land and 'feedlot' farming techniques in the USA potentially produce local problems of high nutrient runoff caused by heavy rain in summer or snow-melt in spring. Up to 28 kg N/ha and 5 kg P/ha were reported from experiments with manure application to arable in Sweden, mostly lost during snow-melt (Brink, 1975), but between 30 and 1300 kg P/ha in runoff from a Nebraska, USA feedlot (Cooke, 1975).

Another relatively new agricultural source of nutrients, particularly in upland Britain, is fish farming. Fish farms, primarily for salmonids, are point sources, posing potential problems because they are located in areas where nutrient runoff is naturally fairly low and water quality high. Phosphorus output is in the region of 10–20 kg P/tonne fish produced (Solbé, 1987).

2.3.6 Urban and point sources of nutrients

The most intensive land use of all is urbanisation. Development of urban areas, particularly after the industrial revolution last century, concentrating people and their wastes inevitably had to be followed by the development of waste treatment processes. Some of these combined the collection of urban runoff from roofs and paved areas with the collection of water-borne human and industrial wastes; some did not. Combined storm and domestic sewer systems are the traditional system in the UK (Parker and Penning-Rowsell, 1980) but account for only about 10% of sewered communities in the USA (Weibel, 1969). The advantage of combined systems is that storm runoff from the urban areas passes into the sewage treatment plant; the disadvantage is that at times of high flow many sewage treatment plants have storm overflows which discharge directly to watercourses with no treatment or only minimal settlement. With separate systems, however, there is usually no treatment available for urban runoff at any time. Studies on urban runoff in Cincinnati, USA have shown losses of 1 kg P/ha/ann and 9 kg N/ha/ann (Weibel, 1969), similar to figures quoted elsewhere (Owens, 1970; Balmér and Hultman, 1988) and is about two orders of magnitude less than the content of raw sewage.

The history of sewage collection and water-carriage to treatment over the past century are reviewed by Porter *et al.* (1975a) and Parker and Penning-Rowsell (1980). Currently around 95% of households in the UK and 75% in the USA are connected to sewage treatment plants, with virtually 100% connection in urban areas. The principle of sewage treatment is that the capacity of the sewage to pollute a watercourse is reduced by enabling the bacterial oxidation of organic matter derived from excreta (and any industry which may be connected to the system) to take place within the treatment plant rather than within the watercourse; typically around 80% of the organic matter is oxidised in this way (Mason, 1981). In doing so, however, all the major elements from the wastes – such as carbon, nitrogen, phosphorus – are oxidised and those which are soluble drain in high concentration in the effluent from such plants. Thus sewage treatment plants become point sources of high concentrations of nitrogen and phosphorus as soon as they are operational.

Since the 1940s, a further factor which has contributed to the nutrient content of sewage effluents has been the use of detergents (Fig. 2.7). The first detergents, introduced as more efficient than soaps for garment washing in the 1940s, were not broken down in sewage treatment works and caused extensive foaming in receiving watercourses. This was considered an aesthetic and public health problem, so they were progressively replaced by bio-degradable detergents more easily broken down in sewage treatment. Phosphates and poly-phosphates made up as much as 50% of the content of detergents because they considerably enhanced the washing efficiency of the product for several reasons, such as softening and buffering the water, emulsifying oils and grease, and dispersing suspended dirt particles (Deevey and Harkness, 1973). Detergent phosphorus made up between 10 and 20% of phosphorus entering sewage treatment plants in the UK in 1957, but 47–65% in 1971.

Analyses of the total phosphorus content of raw sewage indicated that about 500 g P/capita/ann came from human wastes and 700 g P/capita/ann from soap and detergent sources in the UK (Deevey and Harkness, 1973). These figures are close to the middle of a range of concentrations reviewed by Vollenweider (1968) and compare with 1.5 kg P/capita/ann from wastes and 970 g P/capita/ann from detergent in the USA (Porter *et al.*, 1975b). Between 2.5 and 6 kg N/capita/ann are contained in raw sewage, almost all of which is of physiological origin.

Sewage treatment to secondary level (sedimentation plus bacteriological treatment) (Mason, 1981) removes about 40–50% of the phosphorus in solid sludge (Collingwood, 1978; Deevey and Harkness, 1973; Vollenweider, 1968) and about the same proportion of nitrogen (Anon., 1983). Domestic sewage effluent from secondary treatment processes thus contains in the region of 1 kg P/capita/ann and 3 kg N/capita/ann (Porter, 1975; Deevey and Harkness, 1973; Anon., 1983). These *per capita* figures are similar to the unit area losses of nitrogen and phosphorus in combined-flow sewage effluents of 21–100 kg P/ha/ann and 90–440 kg N/ha/ann for low- and high-density residential areas respectively calculated from data in Weibel (1969).

Most sewage effluents will also contain a proportion of industrial effluents and stormwater runoff from paved urban areas. Industries with particularly high nutrient levels are those processing foodstuffs – such as breweries, dairies, abattoirs, food canneries and sugar refineries. These may have their own treatment plants of discharge to the municipal plant (usually after partial treatment). Various metal finishing processes may also use phosphorus solutions. These contribute to the total nutrient load to sewage treatment works and cannot thus be separated from domestic contribution. Average phosphate concentrations in a range of industrial effluents in the Birmingham, England, urban area (Deevey and Harkness 1973) were about ⅓–½ those of domestic sewage with the exception of an anodising plant, the effluent of which was four times higher. Industrial effluents made up to a third of inflowing sewage at some works and so could raise the inflow phosphorus by a maximum of about 20% in the case of metal finishing.

Many other factors will obviously affect the nutrient input of individual sewage treatment systems, chief among which is the extent of storm overflow events. Combined sewage systems in Michigan, USA had concentrations in storm overflows which were three times higher in total phosphorus and four times higher in total nitrogen (Benzie and Courchaine, 1966). The proportions of urban runoff to sewage, of domestic to industrial dischargers, the rate of water use by consumers will all make each drainage system unique, such that the nutrient input figures quoted above can only be approximate for any specific river catchment.

More isolated human settlements supplied with running water may not be connected to a sewage treatment system, but rely on septic tank disposal, whereby the breakdown of organic matter takes place within the tank and the overflow is dissipated in the soil. Runoff and nutrient loading from such systems is very variable, depending upon such factors as: whether phosphate detergents are permitted, the age and efficiency of the tank, the nature, length and slope of the runoff drainage culverts, the type and depth of soil, the depth of the water table and the proximity and size of the nearest watercourse. In countries like the United States and the Scandinavian countries, where there may be high development of seasonally-used holiday homes close to lake shores or watercourses (Hetling and Sykes, 1973), the portion of the year the property is occupied is obviously also important in terms of annual runoff. Such properties may also contribute nutrients through lawn fertilisation (Ellis and Childs, 1973).

2.4 RELATIVE IMPORTANCE OF DIFFUSE AND POINT SOURCES IN CATCHMENTS

Most nutrient runoff studies have demonstrated that there is, not surprisingly, a considerable mixture of point source and diffuse source inputs into river and lake catchments. The only general conclusion that can be drawn from comparisons of different studies is that, except for heavily urbanised catchments, point sources are usually most important in the supply of phosphorus whereas diffuse sources are more important for the supply of nitrogen. In the river Great Ouse, an arable catchment in lowland eastern England with several large towns, Owens (1970) found that 90% of the phosphorus was derived from point sources but only 30% of the nitrogen. In the catchment of Loch Leven, Scotland, with 99% of the area agricultural and only 1% urban (Holden, 1975), agriculture accounted for 85% of nitrogen but only 15% of phosphorus. In Lough Neagh, Northern Ireland, whose agricultural catchment includes intensive dairy farming and associated process-ing, 54% of phopshorus input was estimated to come from sewage effluent, 40% from agricultural drainage and 6% from creameries (Smith, 1977). In the Fall Creek catchment in rural New York State, USA, with a population of about 12,000 people, Johnson *et al.* (1976) estimated that 35% of phosphorus came from point sources, 20% from agricultural sources and 35% from natural background sources. On a larger scale, Hodges (1973) summarised the total sources of natural and human-generated nitrogen and phosphorus for the USA, showing that human-generated nitrogen was approximately equal to the upper estimate of natural sources. About half of the human load came from agriculture, half from urban sources. Human phosphorus was two to three times the natural loading, with about two-thirds coming from sewage effluents.

Table 2.1 Examples of nutrient runoff measured from different land uses (kg/ha/ann)

Location and land use	Nitrogen	Phosphorus	Reference
Sweden – forest	9–16	‹1	Ahl & Oden (in Ahl, 1975)
Swiss Alps – forest	8	‹0.5	Gachter & Furrer (1972)
Swiss lowland – forest	96	‹0.1	Gachter & Furrer (1972)
Swiss Alps – agriculture	163	7	Gachter & Furrer (1972)
Swiss lowland – agriculture	210	4	Gachter & Furrer (1972)
Great Ouse, UK – agriculture	117	0.6	Owens (1970)
Scotland – agriculture	260	2	Holden (1975)
Scotland – soft fruit	391	5	Harper (1978)
Scotland – urban	352	83	Harper (1978)
USA – forest	20	–	Bormann *et al.* (1968)
– clearcut forest	600	–	Bormann *et al.* (1968)

2.5 GLOBAL ASPECTS OF NUTRIENT RUNOFF

There is still a dearth of information about nutrient runoff outside the north temperate zone, and in the tropics particularly, where many of the land uses and catchment characteristics may be different. Higher, more intensive rainfall in seasonal patterns means that soils erosion is more significant, which could result in greater export of phosphorus bound to soil colloids. Cullen *et al.* (1988) have shown that, in semi-arid areas, unpredictable rainfall and hence runoff results in considerable variation in phosphorus export from year to year, such that mean export coefficients cannot be used. Rather, the pattern of discharge has to be subdivided into classes of discharge intensity and mean export coefficients applied separately to each class. Different land use problems may apply; Malthus and Mitchell (1988) have shown that agricultural practice in New Zealand differs from many other countries in the world in that phosphate fertiliser application is very high with minimal nitrogen application, the latter nutrient coming from fixation by clovers seeded into pastures.

Patterns of fertiliser use are also different in many tropical countries, with generally lower levels of application because of the cost (usually in foreign exchange). Irrigation of intensive agriculture is more frequent, however, potentially leading to greater nutrient runoff than for equivalent crops in the temperate zone. Studies of eutrophication and nutrient losses have generally taken second place to studies of more serious pollution problems (Dejoux, 1988).

Viner *et al.* (1981) reviewed the state of knowledge of nutrient runoff and budget studies for the African continent. They stressed the general lack of knowledge of the continent and emphasised four general differences likely to be of importance in interpreting future studies. These are:

1. The area of arid lands, higher than any other continent, and consequent lack of plant cover which is likely to lead to low runoff quantities of dissolved nitrogen.
2. The ancient nature and weathered rocks of much of the continent which are likely to lead to low quantities of nutrients, particularly phosphorus.
3. The irregular nature of much of the river flows and hence delivery of nutrients (e.g. a study in Morocco where sampling every 3–4 hours showed that 98% of annual phosphate and 74% of nitrogen were delivered in 4 days).

4. The high inorganic suspended load of many rivers, often of finely divided materials which could be biologically active. These could either be large sources of adsorbed nutrients, particularly phosphorus, or large sinks, depending on their state of saturation with nutrients.

Viner *et al.* (1981) quoted a nutrient output study from rural land drainage, for rangeland in Transvaal, South Africa. Total phosphorus losses ranged from 31 to 265 g P/ha/ann, and total nitrogen 22 g–3.33 kg N/ha/ann. Phosphorus losses from a range of catchments in South Africa without point sources were generally lower than those of similar geology in the north temperate zone per unit area, although higher per unit runoff because of the readily eroded soils (Grobler and Silberbauer, 1985b).

Problems of eutrophication which have been most thoroughly investigated have been serious problems arising from influx of effluents from large urban areas (e.g. from the Pretoria area in South Africa (Toerien and Steyn, 1975), and the conurbation of Alexandria in Egypt (Saad, 1980)) where nutrient loading calculations using the runoff values for urban land uses from north temperate studies are possible. A study of phosphorus export from nine southern African lake catchments gave figures of 50–350 g P/ha/ann for non-urban areas and 330 g–1.6 kg P/ha/ann for urban areas (Thornton, quoted in Viner *et al.* (1981)).

2.6 METHODS FOR ESTIMATING THE MAGNITUDE OF NUTRIENT LOSSES FROM CATCHMENTS

Studies evaluating the effects of nutrient loadings on receiving watercourses almost always have to consider a range of land uses within a catchment, each contributing nutrients. The relative magnitude of each has to be evaluated. It is fairly easy to measure point-sources, such as sewage works and industrial effluents, but relatively difficult to measure diffuse sources such as agricultural drainage and intermittent sources such as urban storm drainage and septic tank drainage. Moreover, no source contains a constant concentration of nutrients or is regular in flow so annual estimates of nutrient input to a watercourse are difficult without intensive sampling (Bailey-Watts and Kirika, 1987), which is rarely possible.

Recognition of these problems has led to the development of methods which try to estimate both the likely nutrient loading to a watercourse from a sampling programme that is known to be inadequate or in its formative stages, and also the likely sources of error in these estimates. Early estimates of nutrient runoff per unit area of catchment were based upon measurements of tributary inflows to lakes and extrapolated to the whole catchment (Sawyer, 1947) and upon estimates of population density combined with average nutrient concentration in human wastes (Vollenweider, 1968). The increase in the number and range of nutrient runoff studies in the last three decades has enabled more accurate summaries of likely catchment exports under different kinds of land uses, based upon literature review of the nutrient loss studies reported. A simple model was initially developed by Dillon and Rigler (1975) based upon the measured phosphorus losses of different kinds of land use in a study area in Ontario, Canada, and the principle that these phosphorus 'export coefficients' could be extrapolated to similar land uses elsewhere. Uttormark *et al.* (1974) and others reviewed the literature available from a more extensive range of north-temperate catchment studies and from these compiled a manual (Reckhow *et al.*, 1980) summarising catchment losses of nutrients. These data allow typical export coefficients to be estimated for any catchment whose land uses are known or could be mapped, to estimate a lake's total annual nutrient input. They also suggested that upper and lower limits should be attached to the estimated inputs to reflect the extent to which the investigator thinks the catchment under study matches the real catchments reviewed in the manual. Table 2.2 shows the range of phosphorus export coefficients for major land use categories.

The model developed (Reckhow *et al.*, 1980) eventually leads to an estimate of the loading of phosphorus per unit area of lake surface and of the likely biological consequences arising from the level of

Table 2.2 Phosphorus export coefficients for major land uses (from Reckhow and Simpson, 1980); values in kg/ha/ann except septic tank drainage

	Agriculture	Forest	Rainfall	Urban	Septic tank (kg/capita)
High	3	0.45	0.6	5	1.8
Middle	1	0.25	0.4	0.14	0.7
Low	0.1	0.02	0.15	0.5	0.3

loading. This end use is explained fully in Chapter 3, but the principle of estimating the loadings is explained here:

1. The area of the catchment under study is measured using the best available maps or aerial photographs backed up by field survey if necessary, and the area of each land use category is estimated by the same combination of methods.
2. An export coefficient is chosen for each of the land use categories from the range of literature sources available, using literature results as similar to the study catchment as possible and taking into account all the known characteristics of the catchment (such as whether forested areas are natural or receive fertilisation and the intensity of agricultural use in different land use categories).
3. The export coefficients for each land use category are multiplied by its area to give an annual total phopsphorus export for each category.
4. The contribution of septic tanks based upon the average proportion of year occupied and average number of people per dwelling is calculated.
5. Point source inputs are identified and their impact assessed as volume multiplied by estimated phosphorus content of effluent.
6. The figures for 3, 4 and 5 are added to give the total catchment input of phosphorus to the lake and this figure is used to calculate the phosphorus loading based upon lake surface area and hydraulic renewal.

These steps are discussed in Chapter 5, which deals with the prediction and modelling of nutrients and biological interactions in lakes.

Most of the literature of nutrient losses comes from the North American continent. Nevertheless, the data within them are comparable with the European data reviewed earlier by Vollen-weider (1968). In a comparison of the nutrient loading of three Scottish lakes with contrasting land uses, Harper and Stewart (1987) found very good agreement between the estimates for annual lake areal phosphorus loading based upon published export coefficients, with those based upon mean winter lake concentrations and flushing rate. They found much poorer agreement between estimates for nitrogen loadings, with 'most likely' export coefficient-derived estimates approximately twice as high. They attributed this to the greater variability of nitrogen losses from different agricultural land uses. Thornton and Walmsley (1982) found that phosphorus export coefficients for South African catchments were highly variable, due to less predictable patterns of runoff and soil erosion than experienced in the temperate zone. Nutrient export coefficients are clearly best used within the climatic zone for which they were derived and have a valuable role to play in guiding the planning of eutrophication control strategies.

− 3

The biochemical manifestations of eutrophication

3.1 THE COMPONENTS OF NUTRIENT CYCLES IN AQUATIC SYSTEMS

Natural cycles of nutrients in aquatic systems involve the transfer of different chemical compounds and ions in different quantities through the important biological components (Fig. 1.5). Each nutrient follows different pathways and fluxes through the components which are themselves affected by external changes in nutrient loadings upon the system. The rates of flux are driven by biological processes, which are considered in their own right in Chapter 4, and affected by physicochemical processes such as flow and sedimentation, pH and temperature, and oxygen regimes.

3.1.1 Phosphorus cycles

Phosphorus enters aquatic systems from catchment runoff primarily as particulate forms, adsorbed onto inorganic silt and clay particles. Lesser amounts are in particulate organic form in detritus, and the smallest fraction is dissolved phosphate (Holtan *et al.*, 1988; Golterman, 1973). This is a strongly seasonal process, with 90% or more running off during the winter months or after snow-melt in spring. In the tropics, this seasonality is seen as high inorganic runoff during rainy seasons (Grobler and Silberbauer, 1985b). Point sources of phosphorus often contain a higher proportion of dissolved phosphate, e.g. industrial inflows to Loch Leven, Scotland (Bailey-Watts, 1984).

On entering a lake, many of the inorganic particles and some of the organic ones will be sedimented at rates determined by the degree of mixing in the lake, its depth and retention time, and the mass of the particles (Fig. 3.1). Dissolved phosphate will remain in the lake water column where it joins the existing pool. In temperate winters it may remain in the dissolved form for considerable time, but in summer or in the tropics its residence time is very short.

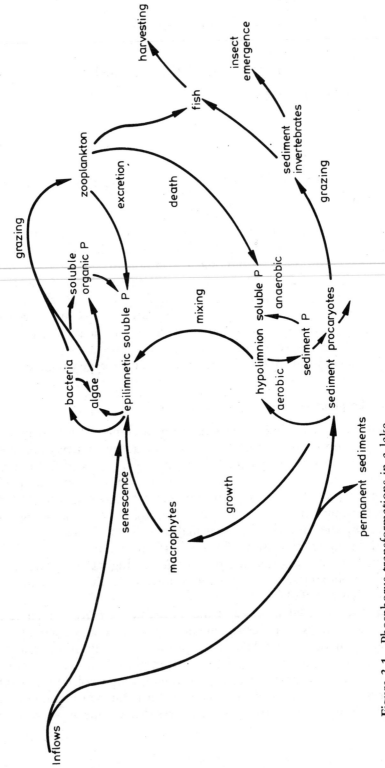

Figure 3.1 Phosphorus transformations in a lake.

Experiments with radioactive ^{32}P tracers have shown that phosphate is rapidly incorporated, within minutes, into organic forms inside algal and bacterial cells (Lean, 1973b). Lean found that in a steady state 98.5% of phosphorus was located in the particulate (algal and bacterial) fraction, 1.16% in a high-molecular-weight colloidal form, 0.21% as dissolved inorganic phosphate and 0.13% as a low-molecular-weight soluble organic form released by the particulates; the uptake of dissolved phosphate by the seston was the fastest process and the release of phosphate from colloidal the slowest, hence rate-limiting step (Lean, 1973a). A part of the colloidal form gradually became unavailable for further particulate uptake.

More detailed analyses of the particulate fraction have shown that it is largely represented by picoplankton (bacteria and small algae) whose turnover rate of phosphorus is very rapid and by phytoplankton (over 30 μm) whose turnover is slower. Phytoplankton have two main mechanisms for enhancing their uptake and use of phosphorus, perhaps necessary because bacteria are more efficient at utilising phosphate (Currie and Kalff, 1984). One is called 'luxury uptake' (Lean and Nalewajko, 1975) whereby cells quickly take up phosphate in excess of immediate needs and store it prior to an increase in population. The other is the ability to secrete phosphatase enzymes externally which cleave inorganic phosphorus from organic molecules. The rate at which this occurs, particularly in relation to chemical hydrolysis of phosphate from organic phosphorus, is not clearly understood, but higher phosphatase activity is generally associated with low levels of phosphate and, conversely, high phosphate levels inhibit phosphatase (Jansson et al., 1988). Bacteria also appear to secrete dissolved phosphate which is taken up, more slowly, by algal cells. The most likely reason for this is that bacteria have the ability to take up phosphorus faster than they can utilise it in growth because they are carbon-limited. Algae secrete organic carbon (Fogg, 1983) which bacteria utilise (Söndergaard et al., 1985); the two populations thus exist in a mutually dependent steady state (Jansson, 1988).

Small protozoa, predominantly flagellates and ciliates, graze the bacteria of the plankton and cause an increase in the cycling of phosphorus both from bacteria and associated organic detritus (Barsdate et al., 1974). Larger zooplankton graze algal and detrital particles, recycling in dissolved form around 50% of ingested phosphorus (Peters, 1975). Phosphorus is thus recycled in two loops within the epilimnion (Azam et al., 1983): between bacteria, algae

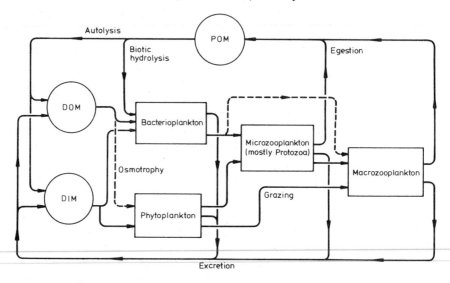

Figure 3.2 Diagrammatic representation of nutrient cycling within the plankton. POM particulate organic matter; DIM dissolved inorganic matter; DOM dissolved organic matter. Modified from Taylor (1982), with permission.

and phosphate with a turnover time of minutes, and between bacteria, algae and grazing zooplankton with a turnover time of hours to days (Fig. 3.2). For each of these loops there is probably tight cycling with only a small percentage loss in sedimentation, but Golterman (1973) has shown that a small loss at each turn of the cycle builds up to a large loss of the total phosphorus pool over the temperate growing season without additional inputs. The grazing–regeneration cycle is likely to be variable, since zooplankton population development often follows that of algae and there are additional sedimentation losses of dead algal cells and zooplankton faeces which may not be fully mineralised before they reach the sediments. Golterman (1976) estimated that 50–80% of mineralisation takes place in the epilimnion before reaching the sediments, but this of course depends upon the degree of mixing and extent of stratification in the lake.

The epilimnion phosphorus fractions may also exchange horizontally with the littoral zone, depending upon the size of the lake and relative proportions of littoral to open water. Early experiments with the addition of radioactive phosphorus to lake water

(Hutchinson and Bowen, 1950) showed appearence of tracer in rooted aquatic plants, probably as a result of turbulent distribution of colloidal organic phosphorus. A phosphorus budget of the littoral zone of Lake Wingra, USA, showed a net export of dissolved phosphorus to the open water approximately double the input of particulates (Adams and Prentki, 1982).

The deep sediments of the lake are the ultimate sink of organic matter produced in the illuminated layers together with the more inorganic silt load from the catchment, and of the phosphorus contained within them. There is some evidence for bacterial decomposition at the epilimnion–hypolimnion boundary in strati-fied lakes (Golterman, 1976), because particles entering the colder, hence denser water of the hypolimnion sink more slowly. The quantitative importance of this is not known, however, and the reactions are probably similar to those which take place on the surface of the sediment.

In the sediments the degree of oxygenation, and hence the redox (reduction–oxidation) potential of the sediment particles, is of over-riding importance to the processes which subsequently take place. (Redox potential is the measure of electrical voltage between two electrodes; it measures the change in oxidation state of many metal ions and some nutrients.) At 25°C and pH 7, oxygenated lake water has a redox potential of about +500 mV and ions are in their stable, oxidised state such as ferric, Fe^{3+}. As oxygen concentration falls different chemical reactions and changes occur at specific redox potential; Fe^{3+}–Fe^{2+} for example occurs at between +200 and +300 mV (Figs. 3.3–3.5). Thus deep stratified lakes with an oxgenated hypolimnion and sediment surface behave differently from shallower stratifying lakes whose hypolimnion becomes progressively deoxygenated through the summer or calm period. Shallow, well mixed lakes without permanent stratification behave differently again, because here there may be temporary periods of anaerobiosis at the sediment–water interface but also periods where the mixing processes of the lake disturb and resuspend the sediment surface particles. Sediments in the littoral region of a lake are likely to remain aerobic and well mixed, although there may be extensive modifications of this scenario by the influence of rooted aquatic macrophytes, considered below.

Aerobic decomposition in sediments proceeds in the same way as epilimnetic decomposition, with microbial oxidisation of the organic carbon sources of detritus for growth which liberates dissolved phosphate in the process. This is temperature-dependent,

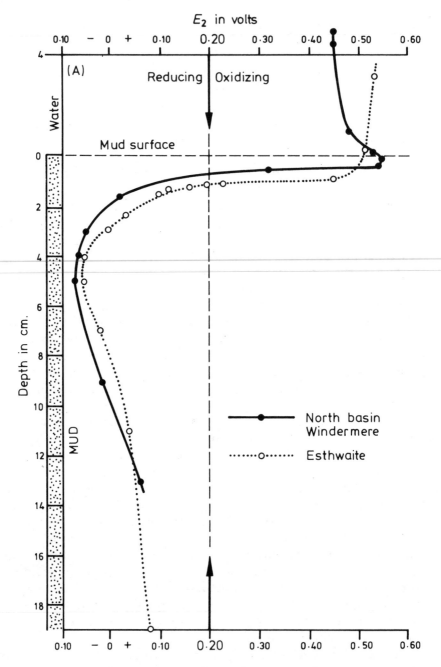

Figure 3.3 The redox potential at the mud-water interface in two English lakes in winter. The redox potential of the mud surface remains well in the oxidised state. Modified from Mortimer (1942), with permission.

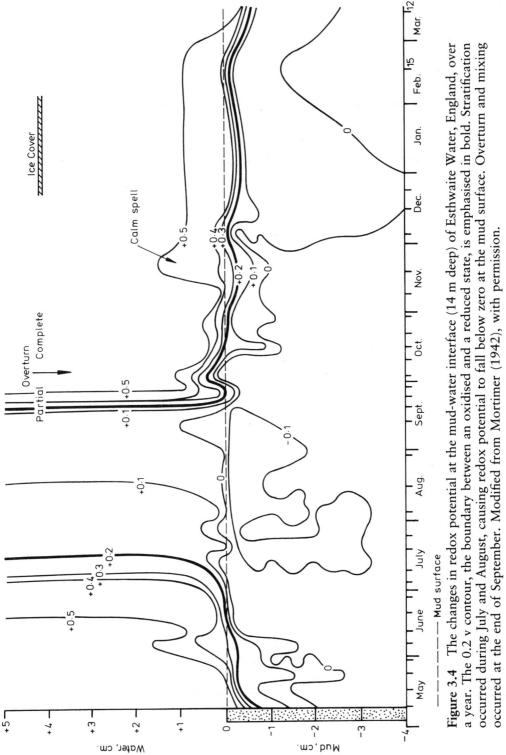

Figure 3.4 The changes in redox potential at the mud-water interface (14 m deep) of Esthwaite Water, England, over a year. The 0.2 v contour, the boundary between an oxidised and a reduced state, is emphasised in bold. Stratification occurred during July and August, causing redox potential to fall below zero at the mud surface. Overturn and mixing occurred at the end of September. Modified from Mortimer (1942), with permission.

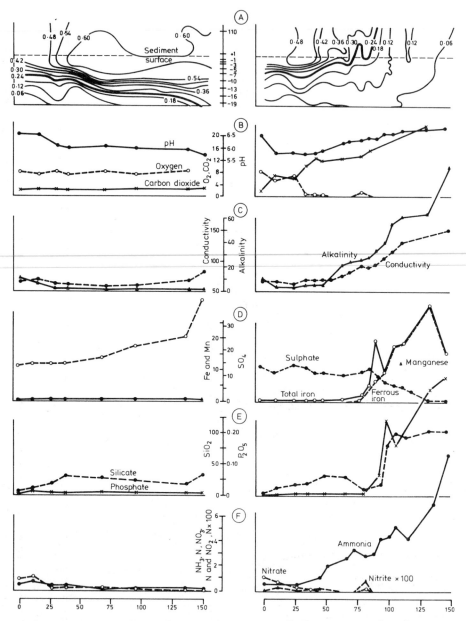

Figure 3.5 Experimental results from two sediment-water tanks, showing the changes in redox potential and concentrations of various ions over a 152-day period. The left hand results show the tank which remained aerated, the right hand results show the tank which was sealed from the atmosphere and became anaerobic after about 30 days. Modified from Mortimer (1941), with permission.

and also regulated by factors such as pH. Released phosphate is then adsorbed chemically by three processes whose relative importance differs in different kinds of lakes. The first is the classic adsorption pathway (Mortimer, 1941, 1942) of adsorption onto Fe^{3+}, probably FeOOH. The second is the complexing of phosphate with calcite ($CaCO_3$) on the surface of calcite particles to form insoluble apatite [$Ca_5(PO_4)_3(OH)$] (Löfgren and Ryding, 1985). The third is adsorption by clay particles (Golterman, 1984). The relative importance of the latter two processes is greater in shallow, hard water lakes. The reverse process, the release of phosphorus into the overlying water, depends upon the relative concentrations in sediment pore water and overlying lake water. The equilibrium between pore water and sediment particle is established rapidly, but rate of exchange between pore water and lake water is slow and normally controls the rate of flux. There may also be a physical barrier formed at the sediment surface by the aggregates and gels of Fe^{3+}-bound P. Additional sedimentation, particularly of mineral silt particles from the catchment, may then progressively reduce the likelihood of phosphorus return. Conversely, return of phosphorus to lake water is more rapid if the sediment is disturbed, by wind-induced mixing or by the burrowing activities of benthic animals.

Under anaerobic conditions two kinds of pathway, biological and chemical, contribute to phosphorus release. The development of anaerobiosis slows down overall microbial activity and hence re-incorporation of mineralised phosphate into new bacterial biomass. Anaerobic bacterial respiration then begins, primarily the process which uses sulphate in place of oxygen (the bacterium *Desulphovibrio desulphuricans*). Thus, using formate as an example:

$$SO_4^{2-} + 4HCOOH - 4CO_2 + 2H_2S + 2OH-$$

Reduction of ferric hydroxide may then occur by the liberated hydrogen sulphide. Microbial reduction of ferric ions may also occur using Fe^{3+} compounds as an alternative electron acceptor in analagous fashion to denitrification (see below) (Boström *et al.*, 1988). There are also various microbial nitrogen transformations and there may ultimately be the production of methane (CH_4) by the complete reduction of organic carbon molecules, using various substrates. It is possible in some sediments for gas production to disturb the upper sediment layers and thus enhance phosphorus release from interstitial water. The chemical reduction of Fe^{3+} to Fe^{2+}, which, being soluble, causes release of the bound PO_4, takes place inevitably as the redox potential is lowered to below +300 mV. Anaerobic release of phosphate may be up to two orders of magnitude greater than aerobic release.

There is thus a complex set of different processes controlling exchange of phosphorus between sediments and water. It can be further regulated by the influence of nitrate, which may buffer the redox potential of the surface sediments at a level high enough to inhibit release of iron-bound phosphorus. The processes of nitrogen cycling in lakes are examined below.

3.1.2 Nitrogen cycles

The nitrogen cycle involves considerably more pathways and reservoirs than phosphorus, because of the variety of oxidation states in which the nitrogen molecule can exist, and in consequence the cycle is more complex and less well understood.

The most stable form is nitrate (NO_3), which may be microbiologically reduced to nitrite (NO_2), although the reverse chemical oxidation rapidly occurs in aerobic conditions, and further reduced to nitrogen gas (N_2). Biological fixation of nitrogen gas by bacteria and cyanobacteria is an important process in soils and lake waters, reducing nitrogen inside the cell to ammonia (NH_3). Ammonia is the most reduced form of inorganic nitrogen, occurring in water partly as the dissolved gas but mostly as the ammonium ion after dissociation of a water molecule (NH_4^+). Various forms of dissolved and particulate organic nitrogen also occur as a consequence of biological growth and decay.

Nitrogen input to lakes in drainage waters is primarily in three dissolved forms. The two inorganic are more important, nitrate and ammonium, in that order. As shown in Chapter 2, nitrate in soils is dissolved in interstitial soil water, so is easily leached by rainfall; ammonium is weakly bound electrostatically so less likely to be leached. Ammonium (which, though toxic, is found in high quantities only in polluted waters) is generally taken up in preference to nitrate by algal and bacterial cells and there is some evidence that it temporarily inhibits the nitrate uptake enzymes. Nitrate is used, more slowly, as ammonium falls. Organic nitrogen is present in inflows in small amounts, and low-molecular-weight compounds such as urea can be utilised directly. There is some evidence for different forms of dissolved organic nitrogen, at least a labile and a refractory class, and for the possibility that labile DON can be adsorbed onto carbonate particles in hard waters serving both as a substrate for microbial colonisation and oxidation and a direct sedimentation mechanism (Manny, 1972). Consequently, the largest fraction of nitrogen measured in lake waters is usually nitrate and the lowest ammonium.

Within the planktonic community there is limited evidence that bacteria and algae compete for nitrogen sources, probably because of the availability of different forms and possible limitation by the rate of supply of phosphorus rather than nitrogen. In most lakes though, ammonia and nitrate progressively decline in the epilimnion through the growing season. There is evidence for rapid recycling of ammonia and for the quantitative importance of ammonia excretion by grazing zooplankton (which excrete up to 70% of N as NH_3), particularly the protozoan and rotifer bacterial grazers (den Oude and Gulati, 1988). Thus the rate of nitrogen cycling in the epilimnion is probably as fast as that of phosphorus.

An additional input of nitrogen to the planktonic food web occurs during periods when nitrogen-fixing cyanobacteria become quantitatively important in the phytoplankton, often in late summer. The ecological conditions leading to cyanobacterial dominance are explored in Chapter 4, but the process is of limited significance in oligotrophic lakes. It becomes increasingly important in eutrophic ones, where fixation may supply proportions in excess of 50% (e.g. Lake George, Uganda (Horne and Viner, 1971) , Lake Erken, Sweden (Granall and Lundgren, 1971)). This occurs as a consequence of nitrogen limitation in the presence of excess phosphorus: a low nitrogen-to-phosphorus ratio. Nitrogen-fixing cyanobacteria species gradually dominate the plankton of a lake and in so doing increase the nitrogen:phosphorus ratio to the Redfield ratio of close to 16. Schindler (1977) experimentally lowered the nitrogen : phosphorus ratio to 5 in added fertiliser in a whole-lake experiment and stimulated the production of cyanobacteria which raised it back to 14.

Mineralisation of decaying algal cells and zooplankton faeces in the epilimnion seems to be high, with 80% breakdown of algal proteins, less than half of which is incorporated into bacterial biomass (Golterman, 1976). As decaying particles sink, the process slows, as the C:N ratio increases and bacterial activity may become nitrogen-limited. Nevertheless, Hall and Jeffries (1984) showed that half of the total nitrification of sedimenting organic material occurred in the hypolimnion of Grasmere, England, and half in the sediments.

In the sediments the number and activity of bacteria is some three to four orders of magnitude greater than in the overlying water (Jones, 1979). Initial decomposition in the sediments may proceed aerobically in the same fashion as in the epilimnion, with release of ammonia where it accumulates in interstitial water. In a similar fashion to phosphate, it is either re-mixed into the epilimnion in shallow lakes or remains in the hypolimnion in stratified lakes; microbial nitrification of ammonia ceases as deoxygenation sets in. A small quantity of ammonia is chemically adsorbed onto sediment particles, the processes of which are unclear (Jones *et al.*, 1982).

Anaerobic processes may come to dominate the sediments even before deoxygenation of the hypolimnion occurs because of the poor rate of diffusion of oxygen into the sediments and its rapid uptake. As soon as the redox potential falls below about $+350\,mV$ following the disappearence of oxygen, nitrate becomes the next available electron acceptor for carbon oxidation and microbial denitrification to N_2 occurs. This process may remove a significant proportion of the lake's nitrogen as gas evolved (Seitzinger, 1988) but in some lakes its rate may be limited by microbial reduction of nitrate back to ammonia (Jones and Simon, 1981). Denitrification occurs in sediments whose overlying waters are not completely anaerobic, such as littoral and river sediments (Jones and Simon, 1981; O'Neill and Holding, 1975), probably because there can be a sufficient oxygen deficit in sediments for the redox potential to be reached.

If nitrate levels in the sediment pore water are not fully depleted by these two processes over the course of the anaerobic period, the redox potential may remain above the level at which Fe^{3+}-bound phosphorus is released (Anderson, 1982). Thus the relationship between nitrate and phosphate concentrations may have an important bearing on the magnitude of releases of both ammonia and phosphate back into overlying waters of lakes.

3.2 THE ULTIMATE SINKS OF INFLOWING NITROGEN AND PHOSPHORUS

Both nutrients are lost from lake systems in outflow processes but in order to persist as a permanent feature of the hydrological cycle, lake outflows have to balance inflows over the longer time-span (years): thus outflow is unimportant as a nutrient sink to a catchment unless it is related to some form of lake management such as deliberate dilution (see Chapter 7). The sediments are a sink

Table 3.1 Percentages of inflowing nitrogen and phosphorus lost from a range of lakes (taken from Seitzinger (1988) with multiple years' data averaged)

Lake name	N input	P input	Percentage N lost by		% P lost by
	(kg/ann)		Denitrification	Sedimentation	Sedimentation
Mirror Lake	227	22	3	27	76
Gardsjon	784	8	42 (combined)		30
Lake Wingra	31357	1277	26	55	94
Kinneret	3.3×10^6	295500	60	9	88
Kul So	18500	1890	20	3	-2
Kvind So	227090	1955	20	8	-4

for both nitrogen and phosphorus, but additionally nitrogen may be permanently lost to the atmosphere if denitrification exceeds nitrogen fixation.

In a range of lakes for which nutrient budgets over a year or more have been calculated (Table 3.1) only very shallow lakes (<4 m mean depth) in Denmark (Anderson, 1971) showed little or no net sediment retention of phosphorus and minimal retention of nitrogen. This was probably because wind mixing of the upper layers of sediment caused the water column processes to dominate the sediment processes; high pH values of up to 10.5 as a consequence of algal photosynthesis would prevent apatite formation and clay colloid binding, and sediment disturbance would recirculate ammonia. Denitrification occurred in all lakes at average rates varying from 5 to 53% of inflowing nitrogen. In other lakes studies, phosphorus sedimentation ranged from 25 to 87% of inputs. The total nitrogen retention was from 10 to 84%, with denitrification, where calculated, ranging from 3 to 62%.

3.3 CHANGES IN THE CYCLES WHICH OCCUR AS A CONSEQUENCE OF ENHANCED NUTRIENT INPUTS

The lakes in the comparison above range from highly eutrophic to moderately oligotrophic, so there is little that can be concluded about the relative importance of nutrient sinks over time scales of more than a year. These are more likely to be controlled by the range of morphometric factors unique to each lake, and principles governing retention will only come from a more widespread comparison of larger data sets. Differences in nutrient loading rather affect the relative rates of certain parts of each nutrient cycle. Those parts where differences with enrichment can be seen are nitrogen fixation, changes in the nitrogen-phosphorus ratios, and the importance of nutrient return from the sediments, or 'internal loading'.

The most clearly demonstrated difference is the relative importance of nitrogen fixation in the nitrogen cycle. Howarth *et al.* (1988a), reviewing data on 16 lakes, showed that fixation only accounts for <1% of total nitrogen input to oligotrophic and mesotrophic lakes, but between 6 and 82% of input to eutrophic lakes. This is because human influences on catchment land uses generally reduce the nitrogen:phosphorus ratio of lake inflows below 15 through enhanced phosphorus sources leading to the development of nitrogen-fixing cyanobacteria (Table 3.2).

Differences in nutrient loading, particularly the relative inputs of nitrogen and phosphorus, which differ markedly between different kinds of land use (Chapter 2), affect the nitrogen:phosphorus ratios in lakes. An 'ideal' ratio, that equivalent to the ratio required for algal growth would be 16:1 (Redfield, 1934) although different species of algae show a range of optimum growth ratios of nitrogen, phosphorus and carbon. The biomass of algae is generally phosphorus-limited in all lakes, even at N:P ratios as low as 5 (Chiaudani and Vighi, 1974). In oligotrophic lakes of the Canadian Shield, N:P ratios in inflows are around 20, whereas in outflows they approach 30 and in lake retention, 16. Thus, phosphorus is more tightly conserved than nitrogen (Schindler, 1976).

In the range of whole-lake fertilisation experiments conducted in these lakes, additions of low ratio combinations of nitrogen and carbon with phosphorus were always followed by an increase in the ratios as a result of algal-mediated uptake from the atmosphere (nitrogen fixation or carbon dioxide diffusion across concentration gradients). These experimental results help to explain the observed distribution of nitrogen fixation between lakes of different trophic status. In the experimentally enriched lakes, suspended material retained or lowered its N:P and C:P ratios but dissolved components showed dramatic increases in the hypolimnion (e.g. N:P up to around 300), indicating that both N and C were released into solution from sedimenting seston but P was conserved by rapid bacterial uptake (Schindler, 1975). Such increases were not observed in oligotrophic lakes. Both nitrogen and carbon were then

Table 3.2 The relative importance of nitrogen fixation in lakes of different trophic status (simplified from Howarth *et al.* (1988b))

Lake name	Nitrogen fixation as a percentage of total input
Oligotrophic	
Superior, Huron, Michigan	0.02
Mesotrophic	
Lake Washington	0.3
Windermere	0.3
Eutrophic	
Lake Mendota	7
Lake 226 (fertilised)	22
Lake George (Uganda)	65
Lake Erken	82

susceptible to outflow losses as well as losses to the atmosphere by diffusion and denitrification. In the recovery phase of a lake after fertilisation had ceased (Levine and Schindler, 1989), similar processes were recorded over a longer time span, with phosphorus ultimately sedimented, but nitrogen and carbon lost to outflows and atmosphere. Throughout the periods of enrichment by different ratios of fertilisation and of recovery, algal biomass always followed lake water phosphorus levels but not nitrogen or carbon. These conclusions suggest that eutrophic lakes will tend to show increased soluble nitrogen pools relative to phosphorus as the nitrogen is less efficiently recycled and algal biomass remains phosphorus-limited at the level sustained by phosphorus inputs. They should show reduced N:P particulate ratios, for at least the earlier part of the growing season. A study of 15 lakes in Sweden (Fosberg *et al.*, 1978) did show lower N:P ratios with increasing chlorophyll content.

For both nitrogen and phosphorus, the oxygenation state of the upper sediment layers plays a central role in the extent to which organic-bound nutrients are recycled into the photic zone. Anaerobic conditions are more likely to occur, even in shallow lakes with temporary stratification, as the rain of organic detritus on the sediments increases as a consequence of increased nutrient-mediated productivity.

In deeper stratified lakes the main changes which take place with eutrophication are an increase in the measurable pools of soluble nutrients, particularly during the quiescent season and, in the hypolimnion during stratification. Thus dissolved phosphate and nitrate levels increase during winter but in the epilimnion may be depleted as much as in oligotrophic lakes during the growing season by supporting higher planktonic biomass (particularly phosphorus) and by atmospheric loss (nitrogen). The progressive decomposition of this biomass in the hypolimnion can result in elevated levels of ammonia and phosphate together with silicate released from diatom frustules which are recycled into the whole lake mass at overturn.

Complete deoxygenation of the hypolimnion may result in the release of larger amounts of phosphorus from chemical reduction and ammonia from anaerobic mineralisation together with reduced (and therefore soluble) iron and manganese. Oligotrophic lakes with only a recent history of enrichment, such as the lake 227 in the Canadian Experimental Lakes Area, did not show any significant phosphorus return from sediments however, even under anaerobic conditions. Schindler attributed this to the low amounts of phosphorus in the upper mixed layer of sediment, 6 cm thick, which represented 60 years of nutrient-poor sedimentation prior to only two years of enrichment (Schindler, 1976). In these lakes, mixing of the top few centimetres' sediment was probably caused by autumnal overturn disturbance and by the mixing of benthic animals. The increased phosphorus loading of the experimental period was 'diluted' and thus little was returned even under anaerobic conditions. Shallow eutrophic lakes in Alberta showed mixing to a depth of 16 cm (Reynoldson and Hamilton, 1982) which regulated phopshorus release.

Lakes which have been eutrophic for longer periods would, on this explanation, be expected to show progressively greater phosphorus return from internal loading because their sediments are saturated. Observations on the restoration of lakes by reduction of their external phosphorus loadings have demonstrated this. The mesotrophic Lake Sammanish, in the USA, failed to show the predicted response to one-third reduction of nutrient loading in 1968 but began recovery during the late 1970s (Welch *et al.*, 1986) following reduced internal loadings. Lake Norrviken, Sweden, a lake enriched from urban sources for many years (Ahlgren, 1972) showed only small changes after diversion of external sources of phosphorus because the internal loading from sediments very nearly matched the annual sediment deposition. In the shallow Lake

Vallentunasjön, Sweden, external phosphorus loading has been reduced by 92% but internal loading now exceeds external loading during summer periods by 3–4 times, with periods of both gradual aerobic and rapid anaerobic microbial release (Boström *et al.*, 1985).

In shallower, intermittently stratified eutrophic lakes, mixing may return phosphorus and other nutrients back into the upper layers more frequently during the growing season, stimulating further phytoplankton growth. An expression of the 'stirring capacity' of short-term winds (a few days) in shallow Swedish lakes showed good agreement with chlorophyll increases (Ryding and Fosberg, 1982). In extreme cases, such as shallow enriched prairie lakes in North America (Barica, 1974), this may result in a series of oscillations, with collapse of algal blooms through nutrient depletion followed by deoxygenation, nutrient regeneration, and renewed algal growth one to two weeks later. The mechanisms of this return may be aided by the activities of invertebrates, which reach very high densities in eutrophic productive shallow lakes if sediment deoxygenation is not extreme or prolonged in summer. In Polish lakes, the effect on phosphorus release was found to be greater in eutrophic than mesotrophic, and was primarily caused through upward transport of phosphorus-rich interstitial water rather than by defaecation.

3.4 THE IMPORTANCE OF THE LITTORAL ZONE IN NUTRIENT CYCLES

The littoral zone will have an increasing importance in overall lake nutrient cycles in shallower lakes where its area and volume become a higher proportion of the whole. Emergent macrophytes growing in the water–land edge obtain nutrients exclusively from sediment, and whilst submerged plants obtain some nutrients from water, sediment sources dominate. The physicochemical conditions inside littoral submerged plant beds can show distinct differences from open water as well as strong vertical and horizontal gradients (of for example oxygen, temperature and pH) because of the calming effect of the plant masses on wind-induced water movements and also the respiratory and photosynthetic activity of the plant biomass. Large amounts of detritus from macrophyte breakdown may accumulate within the littoral zone.

Studies of the role of macrophytes in whole-lake processes have shown conflicting results (Granéli and Solander, 1988). In oligo-trophic lakes with low biomass turnover and oxidised sediment surfaces, internal recycling of phosphorus is reduced and the sediments may act as a net sink. In eutrophic lakes, the redox potential of sediments may fall low enough for sulphide formation and release of Fe-bound phosphorus. High pH levels produced as a result of macrophyte photosynthesis would also lead to release of Fe- and Al-bound phosphorus. However, in hard lakes, the production of calcite by macrophytes may in some instances encourage phosphorus retention by Ca-adsorption and marl formation. Any influence which macrophyte beds have on the open water could be accentuated during the growing season of stratified lakes because horizontal diffusion takes place within the epilimnion, at a time when processes within the hypolimnion return little or no nutrients to the epilimnion. Senescence and decay of macrophytes can have significant effects over a short time-period; Landers (1982) showed this for a large soft-water reservoir in Indiana, USA. Annual senescence in late summer supplied quantities of nitrogen and phosphorus to the open water producing significant increases in chlorophyll content. On an annual basis nitrogen was less significant (2% of total inputs) than phosphorus (18%). The most extensive budget studies reported, on Lake Wingra, USA, also showed a net export of phosphorus during the summer period (Adams and Prentki, 1982).

3.5 SEASONAL PATTERNS OF NITROGEN AND PHOSPHORUS CYCLES IN LAKES

The complexities of the different pathways of nutrient flow mean that lakes show individual characteristics within their seasonal patterns of nutrient cycling. Nevertheless, there are certain general-isations which can be made about lakes of different size and nutrient status.

Almost all lakes show a pattern of nutrient increase related to rainfall and runoff. In temperate lakes this leads to a build up in the lake water progressively over the winter or in early spring after snow-melt, before biological activity begins to utilise the pool of nutrients. In tropical regions this may occur twice a year with seasonal rains, but with little measurable accumulation of soluble nutrients as biological activity is continuously high throughout the year. Whatever a temperate lake's trophic state, its biological activity will deplete the nutrient pool as growth commences in spring, regulated by increasing light and temperature. Oligotrophic lakes are more likely to show steady depletion of all soluble nutrients throughout the growing season, whereas eutrophic lakes often show irregular fluctuations, albeit to lower maxima than in winter, as sudden events such as collapse of algal blooms, or of zooplankton populations, or from mixing-induced sediment re-cycling cause releases of nutrients to occur more rapidly than they can be taken up (Figs 3.6–3.8).

3.6 IMPORTANT FEATURES OF NUTRIENT TRANSFORMATIONS IN RIVERS AND ESTUARIES

The essential cycles of nutrients described above occur in rivers and estuaries with the major difference that water flows and inorganic sediments often dominate the environmental factors which regulate the cycles, and that rooted plant communities, analogous to the littoral zones of lakes, play a more central role. In the middle and upper reaches of rivers planktonic interactions are largely absent; sediment processes (including sediment detritus and invertebrates) dominate, and nutrients are 'spiralled' downstream rather than cycled in one place (Welcomme, 1985). The length of a spiral (the theoretical distance travelled by a nutrient atom between one biological incorporation and the next) is affected by small-scale features of the river channel such as debris dams (Likens, 1984). Natural inputs of nutrients in drainage are low, with the main source being terrestrial leaves which are progressively colonised and decomposed by microorganisms and broken up into finer particles by invertebrate 'shredders'; both processes release dissolved and fine particulate material for downstream uptake and consumption. In the lower reaches flow may be reduced enough for planktonic phytoplankton to develop, but very rarely large enough for larger, slow-growing cyanobacteria so planktonic nitrogen fixation is negligible. Flow is never reduced enough in the main channel(s) to allow permanent sedimentation of organic detritus except where obstacles, such as rooted plants, occur and sediments accumulate, or in the floodplain where lentic and swamp conditions may develop. Turbulence normally ensures oxygenation of the water column and sediment in unpolluted waters so that anaerobic processes are confined to the muds of emergent plant beds. Nevertheless, anaerobic processes such as denitrification are import-

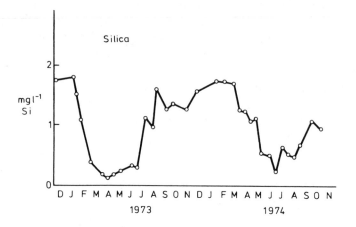

Silica

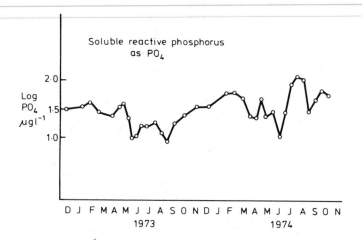

Soluble reactive phosphorus
as PO₄

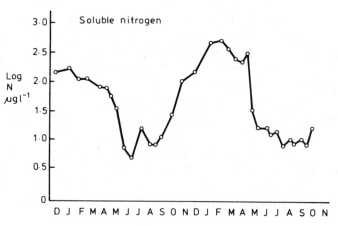

Soluble nitrogen

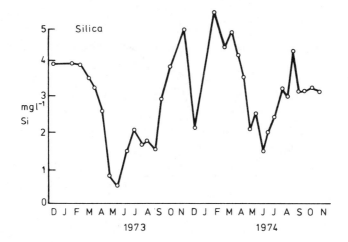

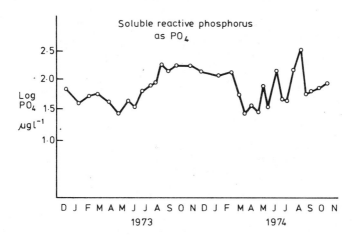

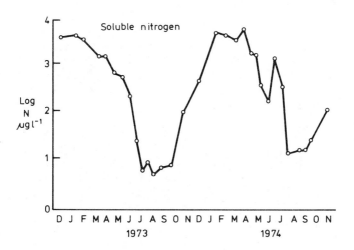

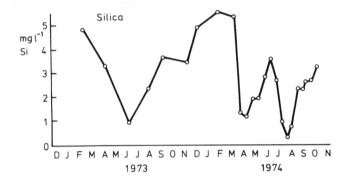

Silica

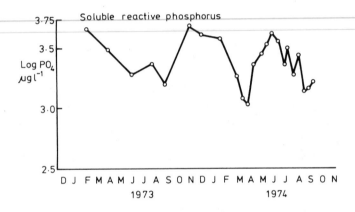

Soluble reactive phosphorus

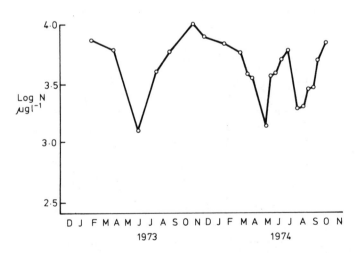

ant, accounting for 50% of incoming nitrogen in some streams studied (Kaushik *et al.*, 1981). Floodplain ecosystems, such as swamp forests or reedswamp, act as sinks for both nitrogen and phosphorus, recycling and retaining nutrients deposited primarily in particulate form during inundation events, with atmospheric exchange (via soil surface nitrogen fixation and denitrification) for nitrogen (Brinson *et al.*, 1983; Yabro, 1983; Howard-Williams, 1985). Detailed knowledge of nutrient cycles in many river systems is lacking, mainly because of the difficulties of using radioactive tracers in an ecosystem where recovery would be difficult.

Nutrient cycles in estuaries are dominated by the complex water mixing and sediment suspension characteristics caused by the interaction of freshwater flow and tidal movements. Denitrification is an important process, with higher loss rates as a percentage of nitrogen input than in rivers and lakes (Seitzinger, 1988), but not as a proportion of sediment nitrogen transformations. The reverse process, nitrogen fixation, is insignificant in most estuaries (exceptions are the Baltic in Europe and the Peel-Harvey in Australia) because planktonic cyanobacteria are restricted to the freshwater component. Rates of benthic nitrogen fixation can locally be very high, for example in cyanobacterial mats associated with the sediment surface of tidal macrophyte beds, but the overall areas covered are low in proportion to total estuarine area (Howarth *et al.*, 1988b). Two reasons postulated for the low significance of estuarine nitrogen fixation are suppression of nitrogenase synthesis by high sediment concentrations of ammonia, which prevents benthic nitrogen fixation, and the low availability of iron and molybdenum in seawater which are essential trace elements for the enzymic process (Howarth *et al.*, 1988a). Consequently, many estuaries have low N:P ratios and production is likely to be nitrogen limited.

Figures 3.6-3.8 Examples of seasonal changes in nutrient concentrations in shallow lakes. Silica and phosphorus show high winter levels from inputs and recycling, with spring depletion by algal growth and slow regeneration through summer, erratic in the case of phosphorus which is associated with fluctuation in algal populations. Nitrate shows a more steady decline from winter peaks, because of additional losses through sediment denitrification. Note that nitrate and phosphate concentrations are on a log scale. Fig. 3.6 Loch of the Lowes (mesotrophic); Fig. 3.7 Balgavies Loch (eutrophic); Fig. 3.8 Forfar Loch (polytrophic with high P inputs all year round from sewage effluent). From Harper (1978).

Phosphorus dynamics are controlled by sediment particle inter-actions, with the greatest reservoir in the benthic sediments. There is evidence that estuaries release significant quantities of dissolved phosphorus from suspended inorganic particles brought in by river transport (Froelich, 1988) but that overall, they accumulate phosphorus in sediments and associated wetlands.

-4

The biological effects of eutrophication

4.1 INTRODUCTION

Increases in the supply and availability of nutrients in water bodies affect the rate of primary production of plants, the magnitude of standing crop biomass achieved and the relative proportion of different species. Nutrient supply is only one environmental factor affecting primary production however, with light availability and temperature being the other important factors. Plant production is then available for animal production (consumers) either directly, as living plant biomass or indirectly after death, as detritus. Thus nutrient effects on any consumer component of an aquatic community are indirect through an alteration in the amount, or relative abundance, and nature (e.g. size or nutritional content) of their food supply, which may alter the balance between competitors for the overall food resources. These effects may occur at every consumer level – herbivores, detritivores and predators.

Nutrient-induced changes in the plant community may also indirectly affect consumers by altering the environmental conditions under which they live, for example by raising the pH in primary production or reducing oxygen concentrations as a consequence of bacterial decay of plant detritus.

4.2 PRODUCTION AND SPECIES CHANGES OF ALGAE AND MACROPHYTES

Plant growth is regulated by light, nutrient supply and temperature, controlling rates of photosynthesis and metabolism and by grazing and senescence (sedimentation in planktonic algae), controlling rates of biomass removal.

4.2.1 Environmental factors affecting algae

Light

Primary production is the reduction of carbon dioxide with the hydrogen atom from water to produce simple sugars and then more complex organic molecules using the energy from sunlight trapped by chlorophyll pigments. Oxygen is a waste product of this process:

$$6CO_2 + 6H_2O \rightarrow C_6H_{12}O_6 + 6O_2$$

A portion of the organic molecules thus synthesised is used in cell respiration, the reverse chemical process, which occurs throughout the 24 hours. The remainder is built up into the other metabolites necessary for new cell construction using the energy released by respiration. These latter processes require an adequate supply of nutrients and all the other essential elements and co-factors. Full details of the processes are contained in standard textbooks such as Golterman (1975) and Raymont (1980) and of techniques for its study (based on measurements of oxygen fluxes, ^{14}C tracers or biomass changes) in Vollenweider *et al.* (1974b).

The total amount of carbon fixed in photosynthesis is referred to as gross production (A), the total fixed minus the respiratory oxidation (R) as net production (N), i.e. the production available for growth or reproduction. Both these latter two processes lead to an increase in biomass (sometimes referred to as standing crop), usually measured as concentration of chlorophyll pigments or cell volume in algae, as dry weight in macrophytes. The level of biomass achieved in any lake is controlled by a range of interacting factors. The rate of increase of new biomass is limited by light and nutrient supply, whilst simultaneous processes removing biomass are senescence and death, grazing by herbivores, parasitism by fungi, or turbulence (removing algae from the photic zone or in lake outlets). It is possible to have high biomass and low production if a situation is reached where, for example, in a dense population of phytoplanktonic cyanobacteria lack of light limits new growth but loss processes are low. Conversely it is possible to have high production with low biomass if, for example a population of diatoms is heavily grazed by zooplankton whose excretion recycles nutrients and whose grazing pressure simultaneously keeps algal biomass low and light penetration high.

The physiological control of photosynthesis by light intensity, varying through the day or with depth, is well understood following the work of Talling on phytoplankton (Talling, 1957a, b). Gross photosynthesis initially shows a linear increase with irradiance, passing a threshold called the compensation point below which respiration matches photosynthesis. Thereafter gross photosynthesis exceeds respiration and net photosynthesis can be measured. Maximum net photosynthesis usually occurs at only about 50% of full sunlight, because a complex set of photo-oxidation reactions occur in the cell when more solar energy is absorbed by chlorophyll than can be utilised down the photosynthetic pathway (Fig. 4.1.).

Light is often above saturation levels at the water surface and shows a logarithmic reduction with depth, so that a typical profile of photosynthesis in relation to depth, for a uniformly mixed population of phytoplankton on a sunny day, shows maximum photosynthesis some way below the surface, up to 2–3 m in a clear lake water. Inhibition due to high light occurs closer to the surface, and decline in proportion to light decline occurs below the maximum. The compensation point is usually found to approximate to the depth of 1% penetration of surface light. Many real-life variations on this theoretical curve of photosynthesis occur, of course (Fig. 4.2). The usual ones are due to uneven distribution of algal cells, the species dominating the phytoplankton (since all have individual light–photosynthesis response curves, the change in light intensity over the day and from one day to the next, and the light-regime history of the cells over the previous few hours). Individual algal cells can show a range of adaptations to both high and to low light levels. In low light conditions pigment content and hence photosynthetic efficiency increases (so-called 'shade' adaptation). A variety of modifications including changes in proportions of different chlorophyll pigments (an increase in carotenoids:chlorophylls) occur in high light (so-called 'sun' adaptation) (Richardson *et al.*, 1983). Shade-adapted cells are easily inhibited by high light, and sun-adapted cells are inefficient at low light, with adaptation to new light conditions taking several hours. Thus rapid mixing events in the water column, such as occur in shallow lakes or can be induced by artificial mixing, may alter the phytoplankton population's photosynthetic rates considerably.

Total photosynthesis below a unit area of water surface has been shown by Talling to be a function of the biomass or density of the population, the irradiance just under the water surface, the irradiance of the beginning of light saturation, the vertical extinction coefficient of the most deeply penetrating light (usually green) light and the highest rate of photosynthesis achieved at light saturation (P_{max}) (Talling, 1957a,b).

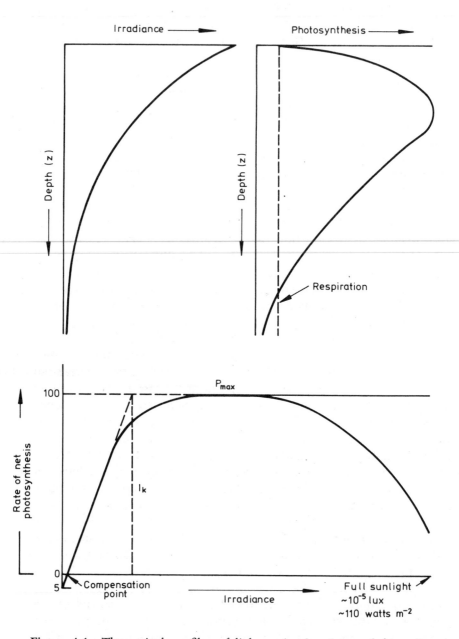

Figure 4.1 Theoretical profiles of light extinction (upper left); primary production (upper right) with depth in the water column; and rate of primary production with irradiance (lower). Modified from Fogg (1980), with permission.

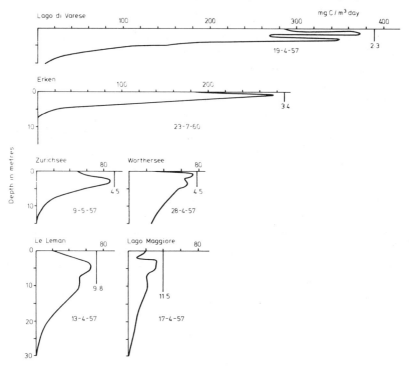

Figure 4.2 Profiles of photosynthesis from lakes of different trophic state. Modified from Golterman (1976), with permission.

Rodhe (1965) showed that production per unit area (P) is a function of biomass (n), the highest rate of photosynthesis achieved at light saturation (P_{max}) and light availability measured as the depth to which 10% of incident light penetrates ($z_{0.1I}$).

$$P = n \, P_{max} \, z_{0.1I}$$

These can be easily measured in the field and production per unit area is a convenient way of comparing lakes with different phostosynthesis–depth profiles.

Temperature

Most biochemical reactions within a cell are temperature-dependent, showing an exponential increase with temperature up to a maximum which varies for each reaction, usually between 25–40°C. Maximal photosynthetic rates for the whole phytoplankton community show such temperature dependence over a seasonal time course in temperate lakes. The rate of increase per 10° rise in temperature (Q_{10}) is usually close to 2. Temperature may also have a more far-reaching effect on photosynthesis and growth by altering the rates of uptake of nutrients and thus affecting the outcome of competition between species.

Nutrient concentration

The effects of nutrient concentrations on production, usually measured as an increase in biomass under non-limiting light and controlled or zero loss rates, have been widely studied in laboratory cultures. A variety of models describing the relationships have been developed which are beyond the scope of the present discussion; recent reviews which deal with these are Ahlgren (1988) and Hecky and Kilham (1988). In outline the rate of increase of biomass with time under no limiting conditions can be described as a function of the starting biomass and the specific growth rate, μ. Growth eventually becomes restricted by the operation of some limiting factor and biomass is held as the 'carrying capacity' of the environment for that factor (K) (Fig. 4.3).

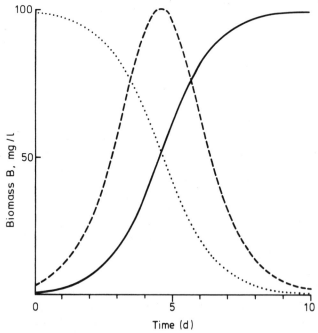

Figure 4.3 Growth of algal biomass with time in an algal culture experiment showing increase in B until K is reached (solid line), the absolute growth rate, rate of change in B with time, dB/dt, (broken line) and the specific growth rate, μ, with time, dB/dt x 1/B (dotted line).

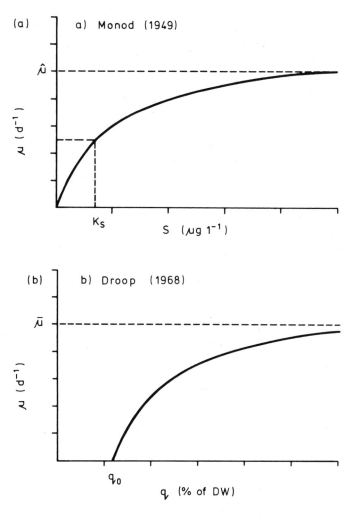

Figure 4.4 Growth rate, μ, as a function of external nutrient concentration (the 'Monod model') (upper) and internal nutrient concentration (the 'Droop model') (lower) from algal culture experiments. Modified from Ahlgren (1988), with permission.

The maximum specific growth rate can be expressed as a function of the external or internal concentration of a single nutrient (Fig. 4.4). Extensive algal culture work has pursued the measurements of μ under different conditions of light, temperature, various nutrients and interactions between these factors (Fig. 4.5) (Ahlgren, 1988; Heyman and Lundgren, 1988). Such experimental

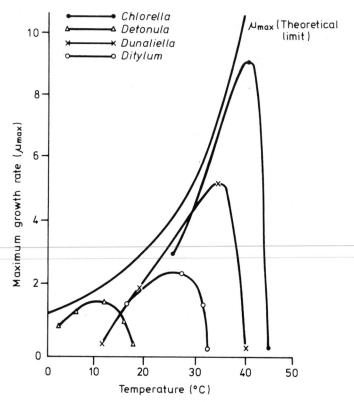

Figure 4.5 The effects of temperature on the maximum rate of growth, $\mu_{m}ax$ for different species of algae. Modified from Welch (1980), with permission.

work has led to investigations of the mechanisms of competitive interactions between algal species which enable the outcome under different environmental factors to be predicted, e.g by Tilman (1977) (Fig. 4.6). Using two species, *Asterionella formosa* and *Cyclotella meneghiniana*, grown in culture with different levels of silica and phosphate Tilman established that each had a characteristic curve of specific growth rate against concentration for each nutrient, and significantly different values of K_s, the concentration which gives half the maximum growth rate. He applied the measured values of μ, K_s and B (biomass) to the growth rate models to predict the outcome of competition in mixed nutrient cultures with different ratios. Where one nutrient was limiting, the species with the lower K_s was predicted to dominate, where both

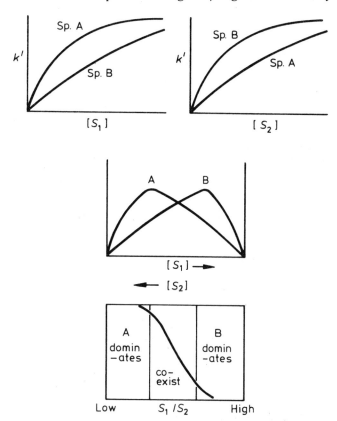

Figure 4.6 The outcome of competition between two species of algae exhibiting different growth rates with concentration changes of a single nutrient (upper graphs); the outcome of competition in cultures with different concentrations of pairs of nutrients (middle) and the prediction of dominance or co-existence based upon nutrient ratios (lower). Modified from Reynolds (1984a), with permission.

were simultaneously limiting, the outcome depended upon the ratio. Where each species was limited by a different nutrient, coexistence occurred. Experimental results supported the model (Fig. 4.7) and also accounted for three-quarters of the variance applied to field observations of the distributions of the algae in a natural gradient of Si:P in Lake Michigan.

The principles apply to other species and to higher taxa such as the outcome of competition between green algae, diatoms and cyanobacteria (Rhee, 1978; Smith, 1983; Tilman *et al.*, 1986). Tilman and colleagues showed that diatoms dominated algal

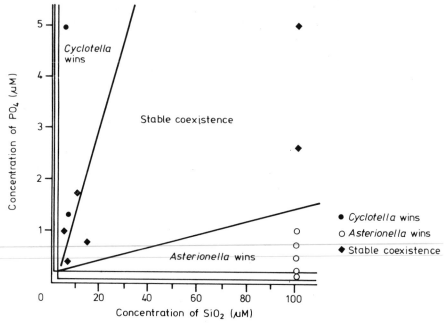

Figure 4.7 Predicted and actual outcome of growth experiments with the diatoms *Asterionella* and *Cyclotella* in cultures with different ratios of phosphorus and silica. Modified from Tilman (1977), with permission.

cultures from Lake Superior at all ratios of N:P at low temperatures, dominance shifting towards greens at intermediate temperatures and ratios and to cyanobacteria at high temperatures and low N:P ratios (Fig. 4.8). High Si:P ratios also favoured diatoms, low Si:P ratios allowed green algae to dominate (Fig. 4.9). The supply of carbon in proportion to other nutrients may also be important, and evidence for species successional changes with pH has been presented (King, 1970; Shapiro, 1973). pH plays a central role in aquatic chemistry and thus affects the uptake of other nutrients in ways not fully understood; for example, high pH decreases the solubility of phosphorus. It is likely that further research will establish that critical ratios of total carbon and phosphorus have effects which can be explained by competition theory.

The change between limitation by one nutrient or another in experiments with pairs of nutrients occurs at different values of μ for different nutrient pairs and is species-specific. Such dual limitation can be either 'interactive' or 'non-interactive' with some experimental evidence for each mechanism. The difference becomes important in relation to maximum biomass which might be

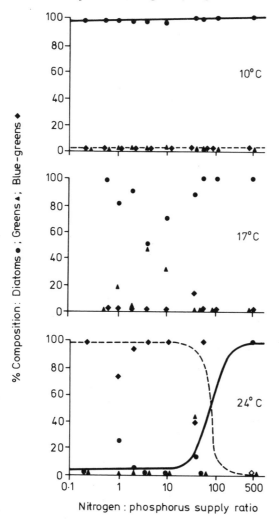

Figure 4.8 The outcome of competition between diatoms, green algae and cyanobacteria in culture from Lake Superior water at different N:P ratios and different temperatures. Modified from Tilman *et al.* (1986), with permission.

achieved in lakes, where evidence suggesting 'interactive' mechanisms implies biomass reductions need to be achieved by lowering of both nitrogen and phosphorus together (Ahlgren, 1988).

Practical applications of these relationships in the field centre around the comparison of measured specific growth rate with μ, and measured biomass with K, under different nutrient conditions.

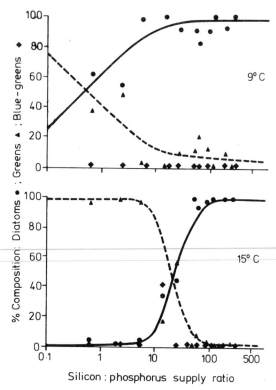

Figure 4.9 The outcome of competition between diatoms, green algae and cyanobacteria in culture from Lake Superior water at different Si:P ratios and different temperatures. Modified from Tilman *et al.* (1986), with permission.

These lead on the one hand to an understanding of how algal species with different μ and K values replace each other as dominants whilst never becoming extinct under the changing conditions within lakes They lead on the other hand to a consideration of the importance of loss processes in determining the extent to which B reaches K.

Under low light conditions, cells are characterised by a high nutrient content but possess a low specific growth rate. When phytoplankton biomass is close to carrying capacity (as in bloom formation) growth rate is limited by nutrients or by light through self-shading but loss processes are low. When light is no longer limiting growth rate increases with temperature, carrying capacity is determined by nutrients but biomass may not reach carrying capacity. Under these conditions (which are the most frequently encountered) the rate of loss of biomass is of prime importance (Heyman and Lundgren, 1988).

Loss processes

Loss processes are due to a variety of factors. Of these, parasitism is of little practical significance in lake management except for short periods at the height of certain phytoplankton species' populations when it may contribute to a decline initiated by other factors such as nutrient depletion. Outflow losses are a straightforward function of the hydraulic renewal time of a lake (or a river in the extreme case) and whilst they will vary seasonally and may retard the growth of populations in temperate lakes at the beginning of the growing season, they can be predicted from a knowledge of runoff and are not normally a major influence on phytoplankton biomass in most lakes.

The main loss processes which may control phytoplankton succession and biomass are sedimentation out of the euphotic zone, which is determined by the degree of turbulent mixing of the water column, and grazing by zooplankton.

Sedimentation and water column mixing

This term is used to cover a suite of processes caused by environmental stress which eventually lead to cell sinking. Most phytoplankton cells can regulate their buoyancy in different ways, ranging from active movement with flagella, through changes in lipid content to secretion of gas vacuoles. Maintenance of buoyancy enables a population potentially to accumulate at the depth of optimal light intensity for photosynthesis, providing these limited powers of movement are not overcome by water column mixing. The rate of sedimentation depends upon the shape and motile properties of the cell (which is species-specific, many cells having adaptations to minimise sinking rates (Fogg, 1975)); upon its physiological state (which determines its specific gravity as a result of the proportions of lipids and other metabolites); and water column stability. A full discussion of the physics of algal sedimentation is given by Reynolds (1984a).

The sinking rate of a particle in suspension is described by Stoke's Law as a function of acceleration due to gravity, fluid density, particle density, and the coefficient of viscosity of the fluid. Deviation from predicted conditions by Stoke's Law occur mainly by variation in the 'form resistance' of an individual algal species' cells.

The inevitable occurrence of dead cells in a colony of living cells, for example, alters form resistance.

Sinking rate is an important parameter to know in detailed studies of phytoplankton ecology. In water management however, it is more important to have a measure of the extent to which a lake's condition will favour algal suspension, or sedimentation. Water column stability, a measure of the resistance of the water column to algal sedimentation, can be calculated in two ways. One, calculated from temperature profiles, is the Brunt-Vasala frequency, N_2 (Sephton and Harris, 1984), which is a function of acceleration due to gravity, the density of water at 4°C and the density gradient over a specified depth interval. A high value of N_2 indicates that density gradients are intense and as a result the resistance to mixing and to sedimentation is high.

A simpler expression is the ratio of the epilimnion to the euphotic depth, Z_m/Z_{eu} (Fig. 4.10). The epilimnion is the mixed depth in a stratified lake, the depth from the surface to the top of the thermocline where the vertical distribution of temperature is more or less uniform as a result of wind-induced turbulence. The euphotic depth is the depth to which 1% of surface light penetrates, which is approximately the compensation point of photosynthesis. Algal cells will be circulated throughout the epilimnion except during very calm conditions when their buoyancy mechanisms may overcome the mixing (e.g. during the formation of surface cyanobacterial blooms). If this is less than the euphotic depth, such as during stratified periods, they will be maintained in light conditions adequate for photosynthesis throughout the day. The reduction of Z_m may, in the temperate spring as stratification sets in, increase overall sinking rate particularly for heavy diatom species and result in their disappearance from the population. If Z_m increases however, such as during mixing events, to a depth greater than the euphotic depth, cells will be swept into low-light conditions inadequate for photosynthesis for part of the day and the overall light regime experienced by the cell will fall so that net photosynthesis will be reduced. In extreme conditions net photosynthesis may approach zero, resulting in cell death or at least the replacement of one species population by another whose growth rate is more rapid under these low light conditions. This is the principle of artificial mixing as a control method for some algal populations.

Approaches to measurement of sedimentation rates have included many different kinds of laboratory experimental systems, producing

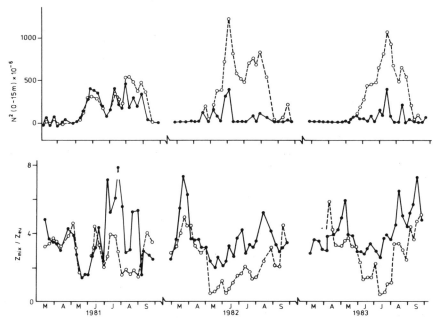

Figure 4.10 An example of the changes in water column stability and in the ratio of mixed depth to euphotic depth in two eutrophic English reservoirs, Staunton Harold, which is artificially mixed (●) and Foremark, which is not (○). From Brierley (1985).

models for sinking rates which are then applied to field conditions (Reynolds, 1984a), direct measurement of sedimenting algae in traps (Bloesch and Burns, 1980; Reynolds *et al.*, 1980) concurrent with detailed recording of the physical conditions of the water column as outlined above, and calculation of loss rates by the difference between specific growth rate and observed biomass changes measured in natural populations (Knoechel and Kalff, 1975).

Estimates of sedimentation rates and their importance as loss processes to phytoplankton populations have shown that they are variable, depending upon physiological state and water column stability. In a small Quebec lake, Lac Hertel, with minimal loss rates (of large algal cells) to zooplankton grazing, Knoechel and Kalff (1975) showed that sedimentation could account for a dominance shift from the diatom *Tabellaria fenestrata*, whose population was restricted by a high sedimentation rate in summer despite a continuing high production rate, to the blue-green *Anabaena planktonica*, whose population was maintained by a lower sinking rate despite a lower growth rate. A further study of the succession of five colonial diatom species in the same lake (Knoechel and Kalff, 1978) showed that cell growth rate for all was determined largely by P_{max}, which despite short-term periods of light inhibition was controlled by nutrient supply. Population fluctuations correlated quite strongly with sedimentation rate. In turn, sedimentation was correlated with death of individual cells within the colonies, which increased their sinking rate. Succession from one species to the next occurred when the net population increase of one species exceeded the net population increase of another. Increased sinking rate as a result of individual cell death rates has also been explained as the cause of colonial diatom decline elsewhere (Jassby and Goldman, 1974), (Lund *et al.*, 1963). Experiments in nutrient-enriched large enclosures (often also called 'mesocosms' or 'limnocorrals') in Blelham Tarn (Reynolds and Wiseman, 1982) showed that the sinking rate of species in different taxa varied markedly; whereas it could account for whole production of the main diatoms *Asterionella* and *Fragilaria*, it accounted for lower proportions of the standing crop of the green algae and less than 4% of the nannoplanktonic organisms *Ankyra*, *Chromulina and Cryptomonas*. Sinking accounted for variable losses of the blue-green *Microcystis*, with resuspension of sediment-ing colonies often occurring through summer until the autumn, when cell lysis following a surface bloom formation was recorded and sedimentation removed the population thereafter. Reduction in the mixed depth, Z_m, with the onset of stratification in spring increased the sinking rate and accounted for the *Asterionella* decline.

Grazing by zooplankton

Loss rates by sedimentation can account for a large part of the losses of large cells but, as shown above, may not be important in the population dynamics of small cells. Grazing effects of filter-feeding zooplankton have been extensively studied in culture and experimental enclosures (Arnold, 1971; Burns, 1969; Frost, 1980; Haney, 1973; Thompson *et al.*, 1982), yielding rates of water filtration and preferred food organisms for zooplankton species and their instars which can be applied to field population measurements (Reynolds, 1984b). These show that the zooplankton community is capable of filtering the volume of the whole lake over short periods at high temperatures and densities in temperate summers. Against this must be weighed the rapid doubling rate of phytoplankton populations (as short as four times per day and in most cases less than once every three days (Lewis, 1974)). Thus in some cases grazing has a negligible effect compared to other loss processes (Lehman and Sandgren, 1985). Comparison of the range of loss processes affecting phytoplankton populations in enclosures in Blelham Tarn, however, showed that grazing losses accounted for significant fractions of the edible algal populations developed in summer, particularly *Asterionella* and *Fragilaria* in July (around 80%), *Chromulina*, *Ankyra*, and *Cryptomonas* in July – August (around 95%), together with possibly a high fraction of a September *Microcystis* population (Reynolds *et al.*, 1982). These and other results (Porter, 1977; McCauley and Briand, 1979; Lynch and Shapiro, 1981) suggest a differential effect of grazing on small (generally <20 μm) sized algae but not on the larger forms in intensity enough to regulate their growth. This in turn affects algal species succession, allowing larger inedible species to rise to dominance in summer under more intensive grazing which suppresses smaller competitors despite their higher growth rates (Gliwicz, 1975).

Grazing effects may thus reduce phytoplankton growth to the extent that the algal community shows reduced or undetectable effects of nutrient changes (McCauley and Briand, 1979). Overall control by grazers should therefore be shown by a negative correlation between zooplankton and phytoplankton biomass in a range of lakes or enrichment experiments, that is when zooplankton biomass is high phytoplankton biomass should be small. This has been found experimentally in enclosures (McQueen *et al.*, 1986) and for certain periods of the year in lakes with variable phytoplankton and zooplankton biomassses such as Lake Dalnee, USA (Brockson *et al.*, 1970) and Lake Siggeforasjon, Sweden (Heyman and Lundgren, 1988). Other data have shown the reverse, a positive correlation between zooplankton and phytoplankton biomass, notably the literature comparison of McCauley and Kalff, (1981). This suggests either that the data collected were not precise enough to show any correlations – mean values for mixed species algal populations will contain both grazed and ungrazed species – or that zooplankton are regulated by the amount of food available and phytoplankton by some other factor such as light and/or nutrient supply. The curvilinear relationship developed by McCauley and Kalff (1981) shows that zooplankton biomass increases more slowly than phytoplankton biomass with increasing trophic state

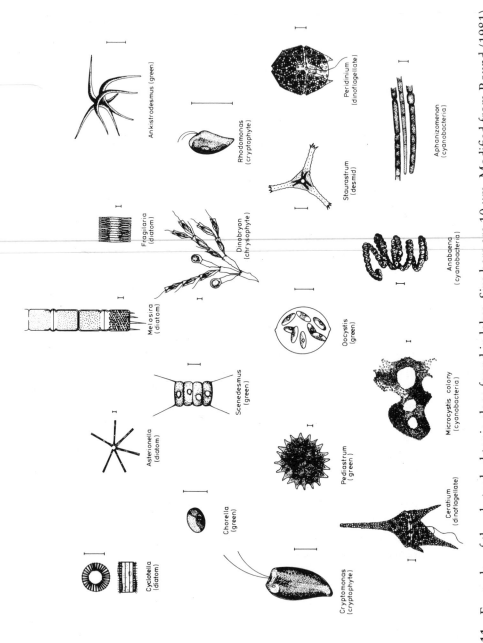

Figure 4.11 Examples of the phytoplanktonic algae found in lakes. Size bars are 10 μm. Modified from Round (1981) and Belcher and Swale (1979), with permission.

which may reflect changing algal species composition. These possibilities are discussed further below.

4.2.2 Effects of eutrophication on phytoplankton dynamics and succession

It has been known for many years that phytoplankton growth in all lakes follows a broadly predictable pattern. Tilman *et al.* (1986) showed experimentally that the most commonly observed taxonomic succession in many lakes, between diatoms in the spring to green algae in summer and to blue-green algae in late summer, can be modelled in laboratory systems by manipulation of nutrient ratios at different temperatures. They found that diatoms dominated cultures at lower temperatures if phosphorus but not silica was limiting (the likely situation in temperate springs); green algae dominated at higher temperatures with moderate or low N:P and Si:P ratios (equivalent to temperate early summers); whilst blue-greens came to dominate at the highest temperatures (24°C) and at low N:P ratios (consistent with their ability to fix nitrogen) which are the typical conditions of temperate high summers. Nutrient levels are one of the most important determinants of the detailed pattern of species change because the same taxa of algae tend to dominate the changes in lakes of similar trophic status but differ between lakes of different trophic status (Reynolds, 1980, 1982) (Fig. 4.11 and Table 4.1).

Table 4.1 Typical phytoplankton species dominating lakes of different trophic states.

Oligotrophic	Mesotrophic	Eutrophic
Staurastrum, Cosmarium Staurodesmus (desmids)	*Staurastrum, Closterium* (desmids)	*Melosira, Asterionella, Stephanodiscus,* (diatoms)
Tabellaria, Cyclotella, Melosira, Rhizoselenia (small diatoms)	*Cyclotella, Stephano-discus, Asterionella* (diatoms)	*Scenedesmus, Eudorina* (green algae)
Dinobryon (chrysophyte)	*Pediastrum, Eudorina* (green algae)	*Aphanizomenon, Microcystis, Anabaena* (cyanobacteria)
	Peridinium, Ceratium (dinoflagellates)	

The initial effects of increased nutrient input to a lake are almost always an increase in biomass of phytoplankton for certain periods, that is, a reflection of the increase in carrying capacity of the environment by increase in supply of a previously limiting factor (Heyman, 1983; Hecky *et al.*, 1988). The relative ratios of photosynthesis to biomass do not appreciably change (Ross and Kalff, 1975; Heyman, 1983). However, this carrying capacity may not always be reached because of the interactive effect of other factors such as nutrient ratios, light, temperature and turbulence which themselves may become limiting. The most extensive synthesis of the interaction of these factors for all algae in lakes has been by Reynolds (Reynolds, 1984b). Oligotrophic lakes are often characterised by a single low biomass peak in late spring–early summer dominated by diatoms; with slightly increasing nutrient status this may be followed by irregular smaller fluctuations through to autumn dominated by desmids and dinoflagellates. Mesotrophic lakes usually show two diatom-dominated biomass peaks, one early and one late; with increasing nutrient status the second peak often matches or exceeds the first. Eutrophic lakes show higher biomass levels maintained for a more continuous period from spring through to autumn with irregular fluctuations as species succeed one another and an increasing importance of cyanobacteria in the later peaks. Highly eutrophic (or hypertrophic) lakes show biomass levels which at their maximum often create light-limiting conditions, with many irregular fluctuations through-out the growing period and an increase in importance of small green and diatom algae. Tropical lakes often have relatively unchanging biomass levels with only small fluctuations throughout the year. The pattern of biomass change with increasing trophic state is shown diagrammatically in Fig. 4.12, and for three shallow lakes in Fig. 4.13. The changes in dominant taxa through the season are shown in a comparison of two oligotrophic lakes with two meso-eutrophic lakes in Figs. 4.14 and 4.15.

Reynolds identified 19 assemblages of species which dominate the seasonal patterns of lakes, divided broadly into spring, summer, autumn and seasonally widespread categories (Fig. 4.16). Analysis of the common physiological characteristics of the species members of the assemblages allowed Reynolds to divide them into three groups (Table 4. 2).

With this outline pattern Reynolds developed conceptual models of the main forces driving the successional patterns. These he termed 'autogenic' if they were directional and slow, such as

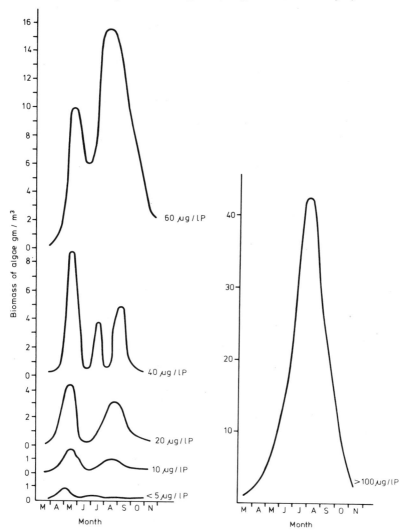

Figure 4.12 Diagrammatic representation of phytoplankton biomass changes in lakes with increasing phosphorus concentration. Modified from Trifonova (1988), with permission.

incident radiation, or determined by the phytoplankton themselves, such as changes in N:P ratios caused by phytoplankton uptake, and 'allogenic' if they were determined by unpredictable irregular outside forces, such as the development of water column stability and its breakdown by mixing. He supposed that the forces operate at two hierarchical levels within a water body. In a temperate lake

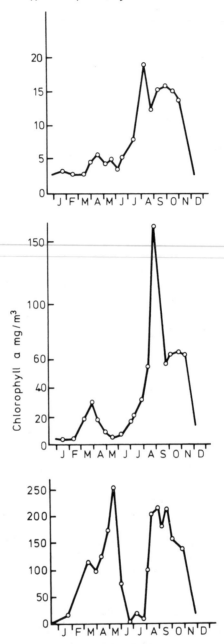

Figure 4.13 Biomass as chlorophyll 'a' in three shallow lakes of increasing trophic state in Scotland. Upper, Loch of the Lowes; middle, Balgavies Loch; lower, Forfar Loch. See Figs 3.6-3.8 for phosphorus concentrations.

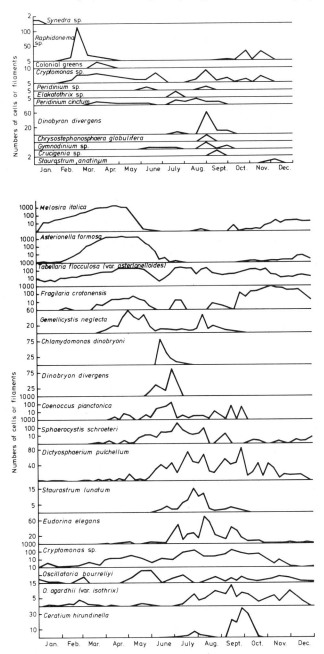

Figure 4.14 Seasonal succession of algae in two oligotrophic lakes of the English Lake District, Buttermere (left) and Windermere (right). Modified from Macan (1970), with permission.

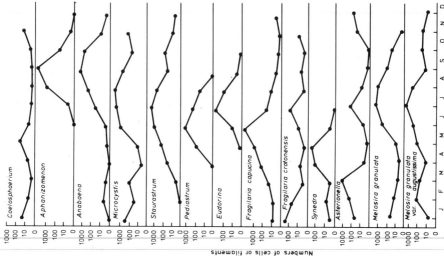

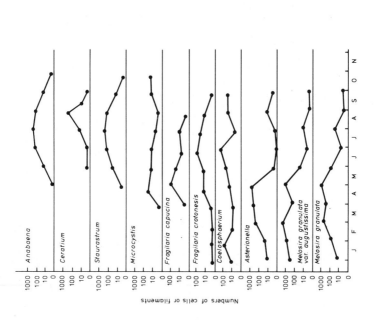

Figure 4.15 Seasonal succession of algae in two meso-eutrophic lakes in Scotland; the Loch of the Lowes (left) and Balgavies (right). From Harper (1978).

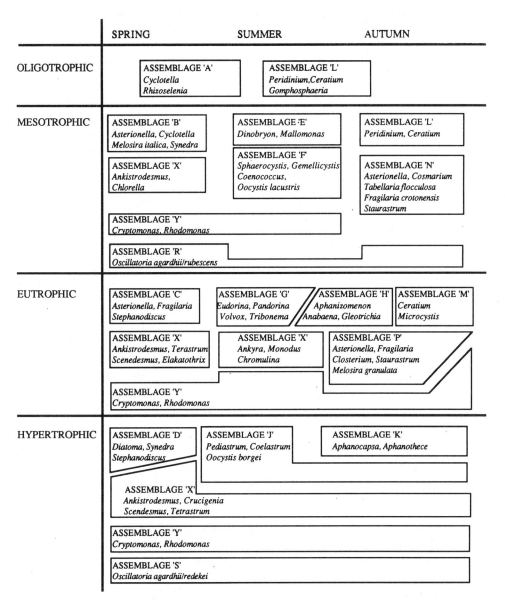

Figure 4.16 Reynolds' 'Assemblages' of phytoplanktonic algae characterising lakes of different trophic state and different seasonal occurrence. Modified from Reynolds (1984b), with permission.

Table 4.2 Reynold's 'Assemblages' of temperate freshwater phytoplankton arranged in three 'groups' according to their approximate tolerances of environmental conditions and their environmental physiologies. From Reynolds (1984b)

Group	Assemblages	Preferred physical conditions	Preferred chemical conditions	Environmental physiologies
I	X, Y	Mixed, low stability, high $Z_{mix}:Z_{eu}$	Lower pH (< 9) higher nutrients	High photosynthetic efficiency, high surface/volume, 'r' strategists, susceptible to grazing
II	A,B,C,D, J,N,P,R,S	Stratified, high stability, low $Z_{mix}:Z_{eu}$	Moderate pH, low to moderate nutrients	Usually high photosynthetic efficiency, moderate surface/volume 'r'–'K' strategists, moderate susceptibility to grazing
III	E,F,G,H, L,M	Tolerant of wide range of stability and $Z_{mix}:Z_{eu}$	High pH (> 9) moderate to high nutrients	Low photosynthetic efficiency, low surface/volume, slower growing 'K' strategists, resistant to grazing

the autogenic forces driving algal change move in importance from predominantly physical factors at the beginning of the growing season (increasing temperature and light penetration, decreasing degree of mixing) through to chemical factors (lower nutrient levels and changing ratios) in the spring–summer and on to biotic factors (increasing impact of grazing and parasitism and allelopathic effects) in late summer.

Assemblages of algae within groups I and II contain higher surface area:volume ratios either by being small and spherical (I) or larger and variably shaped (II) and thus both have higher rates of nutrient uptake, photosynthesis and growth but greater susceptibility to grazing than the assemblages of group III. The latter are larger, with low metabolism sensitive to lower temperatures but less susceptible to grazing. Groups I and II are better able to tolerate well-mixed water columns in spring and autumn, group I can tolerate limited stratification if the euphotic depth is less than the mixed depth. Both are susceptible to nutrient depletion and loss processes (sedimentation/grazing). Group III are large, slower growing but buoyancy regulating and hence not susceptible to loss processes and can withstand high temperatures and light. The autogenic processes generally favour overall succession from groups II in spring to III and back to II again in autumn with group I in intermediate stages between II and III. Within the groups, assemblages change by finer shifts in the same patterns of resource requirements (Fig. 4.17).

Allogenic factors reverse the succession, usually for short periods such as mixing events in shallow lakes in summer. Dominance of one factor such as mixing in most rivers and some exposed shallow lakes, or continuous non-limiting nutrient supply in hypertrophic lakes, or continuous high temperature and irradiance in tropical lakes, tends to suppress the gross succession between groups and leads to finer succession between assemblages.

The ecological basis of these classifications is a development of the concept of 'r' and 'K' selection first proposed by McArthur and Wilson based on the logistic growth equation, whereby 'r' is the intrinsic rate of growth (or maximum growth rate, μ, for algae) and 'K' is the carrying capacity of the environment (MacArthur and Wilson, 1967). Thus 'r' selected species are those which maximise growth rates and whose populations may fluctuate considerably; 'K' selected species are those which adopt a variety of physiological and ecological strategies for maintaining their populations around the carrying capacity. The adaptations of a range of species from any taxonomic or ecological grouping (not only in the aquatic environment) can be compared on a continuum of r–K selection. Thus groups I and II algae of Reynolds' classification can be viewed as possessing a range of 'r' strategies compared with group III algae which display more 'K' strategies: 'r' strategists are more likely to grow rapidly in favourable conditions and decline rapidly when these conditions change; 'K' strategists are likely to grow more slowly over a range of environmental conditions with physiological adaptations which enable them to maintain growth and biomasss. This analysis of strategies has been used to explain

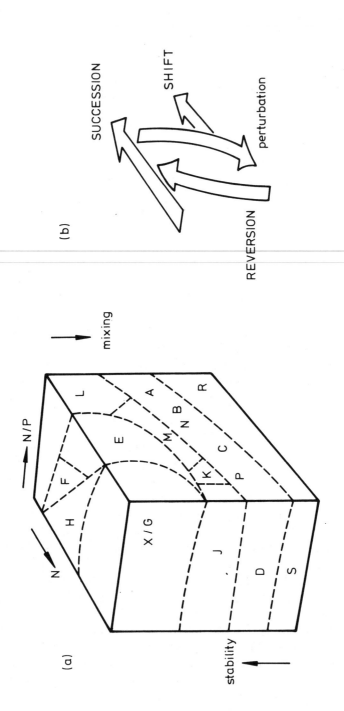

Figure 4.17 A hypothetical matrix showing seasonal succession in a temperate lake according to nutrient staus and turbulence. Modified from Reynolds (1984b), with permission.

algal succession both in lakes and the ocean (Sommer, 1981; Kilham and Hecky, 1988).

This broad overview and explanation of algal succession, though it may lack detail, is playing a very important role in highlighting the gaps in our detailed understanding of processes and formulating hypotheses for their explanation. It encompasses detailed explanations of certain events, such as detailed physiological responses of cyanobacterial pigments (Paerl *et al.*, 1983). It also encompasses more general syntheses of patterns such as the decrease in relative proportion of nannoplankton:larger phytoplankton biomass in lakes with increasing trophy (Watson and Kalff, 1981).

4.2.3 The explanations for dominance of cyanobacteria in phytoplankton succession

When algae reach densities close to the carrying capacity of lakes, rivers and coastal waters (known as bloom formation) they have adverse biological, physical and economic effects. These are discussed in more detail in Chapter 5. The algal taxa responsible for such blooms in freshwater are usually cyanobacteria of Reynolds' group III because their biomass persists for the reasons outlined above. Bloom forming species of similar physiological characteristics but of different taxa occur in marine waters (Paerl, 1988). More research has been carried out on the whole suite of likely causes for blue-green dominance (reviewed by Steinberg and Hartmann (1988) and Shapiro (1990a)) because of their economic consequences. An outline of these causes provides a useful illustration of the detailed processes driving algal succession (Figs. 4.18–4.20).

In deep, stratified lakes with nutrient levels in the mesotrophic range or above (total P > 15–20 µg/l) cyanobacteria slowly build up high populations after more 'r'-selected species have waxed and waned. They do so through growth at lower phosphorus levels (including the ability to store phosphorus in excess of immediate needs) and lower N:P ratios (through the nitrogen-fixing ability of some species). They are more competitive for low nutrients at elevated temperatures (as summer progresses) and more efficient at lower light levels (shading by other phytoplankton). They preferentially utilise CO_2 as a carbon source and are more efficient than other phytoplankton taxa at obtaining this from low concentrations (which prevail at the elevated pH often reached in

(a) Deep stratifying *Oscillatoria*

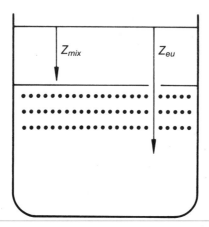

(b) Shallow mixed *Oscillatoria*

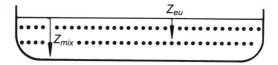

(c) Hydraulically variable *Microcystis*

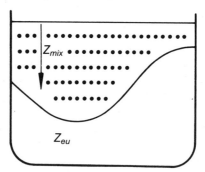

Figure 4.18 An illustration of the way in which different taxa of cyanobacteria respond to the interaction of mixed depth and euphotic depth in shallow and stratified lakes. From Anon. (1990), with permission.

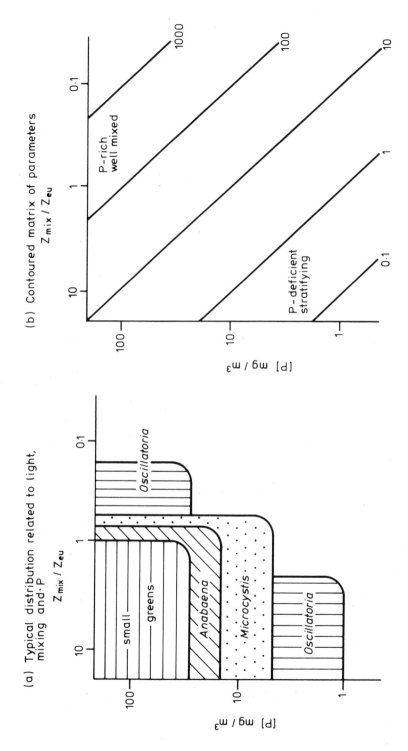

(a) Typical distribution related to light, mixing and P

(b) Contoured matrix of parameters

Figure 4.19 Diagrammatic representation of dominant algal taxa in lakes as a relationship between phosphorus concentration and the mixed depth/euphotic depth relationship (a) and the contours of the ratio $P:Z_m/Z_{eu}$ (b). Modified from Anon. (1990), with permission.

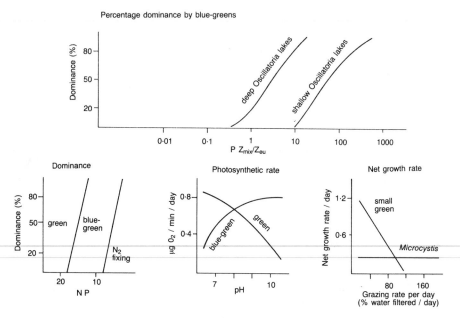

Figure 4.20 Diagrammatic explanations of the dominance changes of algal taxa in lakes. Upper graph shows the changeover of *Oscillatoria* spp. dominance in both deep and shallow lakes at different $P:Z_m/Z_{eu}$ ratios. The lower three graphs explain the outcome of competition between the other algal taxa in the region between the two *Oscillatoria* plots in terms of N:P ratios, photosynthetic rate at differing pH, and net growth rate in relation to zooplankton grazing rate. Modified from Anon. (1990), with permission.

lakes in summer as a result of intense photosynthesis by other phytoplankton). They may also obtain it directly from the atmospheric pool by bouyancy regulation allowing them to float at the water surface. Protection from light damage at the water surface is provided by reverse migration (over a diurnal cycle) and by the development of carotenoid pigments which protect them from UV damage and enhance photsynthetic efficiency. Once their populations have reached high biomass, they effectively shade and may have allelopathic effects upon other competitors, have low loss rates through sedimentation, and are difficult for zooplankton to graze extensively.

In more turbulent environments, ranging from shallow, well-mixed lakes to rivers, the suite of cyanobacterial adaptations is less effective. If Z_m is less than Z_{eu}, buoyancy and low-light adaptations may still be effective in maintaining growth, with some cyanobacterial species such as *Oscillatoria* spp. better adapted to turbulence than others even if Z_m exceeds Z_{eu}. If Z_m consistently exceeds Z_{eu}, then these conditions will tend to enhance nutrient levels by circulation and reduce pH by lowering overall photosynthesis, leading to competitive dominance by other slow-growing turbulent-adapted species such as large diatoms. If the pattern of turbulence is irregular then all the slower growing 'K' strategists may be unable to adapt their light and nutrient gathering physiologies rapidly enough, leading to dominance by 'r' strategists. In fully mixed environments, such as nutrient-rich rivers, 'r' strategists dominate even though cyanobacteria may persistently occur because the turbulent renewal of nutrients and carbon at the cell–water interface favours rapid growth strategies.

4.3 EFFECTS OF EUTROPHICATION UPON MACROPHYTES AND ATTACHED ALGAE

The penetration of light underwater is the primary factor affecting the occurrence and productivity of macrophytes and attached algae. Unlike phytoplankton, however, they also need a substrate for growth and attachment. Spence showed in a survey of lochs in Scotland that depth of colonisation was negatively correlated with euphotic depth (Z_{eu}) but that the relationship became unclear in lakes with high maximum phytoplankton crops and fluctuating Z_{eu}, which produced large annual macrophyte fluctuations (Spence, 1975). Competition with phytoplankton through reduction of light was shown to account for the decline of submerged macrophyte biomass in Lake Wingra, Wisconsin (Jones *et al.*, 1983; Titus *et al.*, 1975). Duarte *et al.* (1985) showed from an analysis of 139 lakes in the literature that biomass and cover of macrophytes were proportional to lake area. Biomass and cover of submerged macrophytes tended to be smaller in large lakes, reflecting the greater effect of open water compared to littoral. Cover was correlated with underwater light at mean lake depth more strongly than was biomass. Biomass could be related, on a finer scale of analysis, to littoral slope (Duarte and Kalff, 1986).

Macrophytes obtain most of their nutrient requirements from sediments (Best and Mantai, 1978; Barko and Smart, 1981) and, as such, do not seem to compete directly with phytoplankton for these, although it is possible they may intercept nutrient inflow and retain nutrients through sediment recycling. Epiphytes obtain their nutrients from the water, but supply of nutrients may be higher within macrophyte beds as a result of leaching and senescence. There is evidence for macrophyte secretion of substances which depress epiphyte colonisation and growth, but the role which this plays in lake communities has not yet been established. Epiphytes have been shown to have negative effects upon macrophyte growth and productivity, particularly in high nutrient levels (Sand-Jensen and Søndergard, 1981). Sand-Jensen and Borum (1984) showed that epiphyte growth inhibited macrophyte depth penetration in nutrient-rich Danish lakes, and that their biomass increased more rapidly than that of phytoplankton with increased nutrient status. This may be through competition for light, but may also be through reduced photosynthetic efficiency of macrophytes at high pH and low CO_2 concentrations compared with epiphytes (Simpson and Eaton, 1986).

Relatively few lakes have had their primary production of phytoplankton, macrophytes and epiphytes studied contemporaneously. Wetzel found in a Californian lake that littoral zone productivity exceeded phytoplankton productivity on an areal basis, with epiphyte production an order of magnitude greater than that of macrophytes. Over the whole lake however, phytoplankton annual production was greater (Wetzel, 1964). In shallow hard-water or littoral-dominated lakes macrophyte productivity may be higher, accounting for 40–50% of annual production (Rich *et al.*, 1971) (Raspopov in Ravera *et al.* (1984)).

Studies which have compared macrophyte distribution in lakes of different trophic state or in the same lakes over time have found a similar pattern of macrophyte change (Phillips *et al.*, 1978; Spence, 1982; Harper, 1986) discounting the often dramatic effect of recent invasion by alien species such as *Myriophyllum spicatum* or *Elodea candensis*. Lachavanne (1982) outlined five stages of macrophyte change in Swiss lakes which have wider applicability to lakes of the temperate zone:

1. Ultra-oligotrophic phase with macrophytes rare.
2. Oligotrophic phase with macrophytes dominated by charophytes colonising to a depth of 25 m.
3. Mesotrophic phase with a notable increase in species richness and abundance, dominance changing from charophytes to *Potamogeton* species.
4. Eutrophic phase with high abundance but reduced diversity and

restriction of depth colonised.
5. Hypertrophic phase with disappearance of submerged species.

The mechanisms for progressive macrophyte decline in all lakes with increasing enrichment are unclear, and a general explanation with several different hypotheses for the events triggering rapid change has been proposed (Scheffer, 1989; Stansfield *et al.*, 1989). Low nutrient status lakes have clear water, little phytoplankton or epiphytes, a relatively deep Z_{eu} and submerged macrophyte beds. They also contain diverse invertebrate and fish populations; these include herbivorous zooplankton and predatory fish which need the macrophytes for shelter from each other. Increase in nutrient inputs may cause increase in macrophyte biomass through 'luxury uptake' and tight recycling over a range of nutrient levels without increase in phytoplankton density. A progressive increase in epiphytes occurs, however, and increase in macrophyte detritus and epiphyte production leads, through invertebrates, to an increase in invertebrate-feeding fish such as cyprinids. These cause turbidity to increase in the water, both through direct disturbance of sediment in benthic feeding and through reduction of zooplankton density which then exerts less effective grazing control of phytoplankton biomass. Reduction of zooplankton density may also have been through other external factors, such as runoff of pesticides which were widely used in the 1960s. Macrophytes die down each winter and re-grow in spring from overwintering turions on the bottom. At a certain level of turbidity this spring re-growth is prevented, and over a short period of only a few years macrophyte populations collapse. With the decline of macrophytes cover for predatory fish and substrate for littoral invertebrates is lost. Both effects cause further increase in cyprinid density and cyprinid grazing pressure on zooplankton. By this stage high phytoplankton crops may cause anoxic sediment conditions, leading to reduction of benthic invertebrate diversity and biomass, further increasing grazing pressure on zooplankton. The end result is a lake with dense phytoplankton biomass and a high population of stunted (food-limited) cyprinid fish but little else.

Emergent macrophytes, which are not in direct competition for nutrients with either algae or submerged plants, may show enhanced growth and biomass over a wide range of nutrient inputs. A comparative study of populations of the reed *Phragmites australis*, growing in three Scottish lochs, which ranged from mesotrophic to hypertrophic (Harper and Stewart, 1987), showed increased growth, biomass and efficiency of solar energy conversion (Ho, 1979), with the latter ranging from 1.3% in the mesotrophic Loch of the Lowes to 2.5% in the eutrophic Balgavies Loch and 7.7% in the hypertrophic Forfar Loch. Under certain conditions of continuous high nutrient input however, decline of emergent macrophytes occurs, and has been reported in several European countries (de Nie, 1987). Several mechanisms interact. Direct effects of nutrient enrichment is a stimulation of growth, greater stem density and acceleration of longitudinal growth; this leads to greater intra-specific competition for light and a weakening of individual stems. Indirect effects of enrichment are the general enhancement of algal growth which leads to increased detritus loadings (and possible deoxygenation) in the reed-bed sediment together with enhancement of epiphyte 'collar' growth on stems. These two may exert a weight upon the stem which lessens its stability under wave action, leading to dieback at the water's edge (Kloetzli, 1982). There is also evidence that elevated nitrogen levels cause reduction in sclerenchyma tissue and loss of stem strength.

4.4 PRODUCTION AND SPECIES CHANGES OF ZOOPLANKTON

4.4.1 Introduction

Zooplankton range in size from a few microns (flagellate protozoa) to a few centimetres (crustacea or insect larvae), with most falling in the size range 50 μm–5 mm. The predominant feeding method is filtering of the range of living and detrital organic particles in the plankton, with larger species feeding raptorially upon large algae or other zooplankton. All are to a greater or lesser extent size-selective, with the range of particle sizes each species can ingest determined by critical dimensions of their food-gathering apparatus, such as the dimensions of the carapace gap and distance between setules of filter-feeding cladocerans (Fig. 4.21).

Most of the zooplankton in freshwaters are members of three main taxa. The phylum protozoa is mainly represented by ciliates and flagellates; these together with the phylum rotifera make up the microplankton up to 500 μm. Crustacea make up most of the macrozooplankton, largely cladocera and the copepoda groups calanoidea and cyclopoidea. These groups have different reproductive strategies which influence the rate of population increase and hence responses to food availability. Protozoa can (and usually do) reproduce by simple fission, with sexual reproduction confined

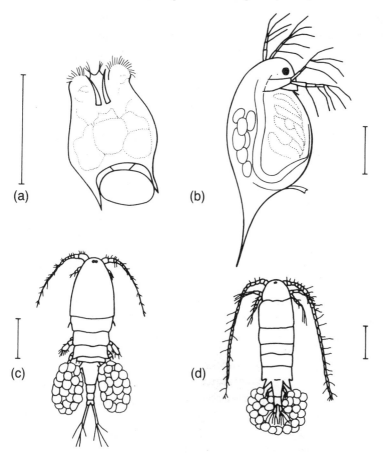

Figure 4.21   Illustrations of planktonic zooplankton in lakes, size bars are 0.5mm. Modified from Moss (1988) with permission. They are (a) Brachionus (rotifer), (b) *Daphnia, Cyclops* and *Diaptomus* (crustacea).

to relatively rare periods of response to adverse conditions such as low temperatures and often involve the production of resting stages. Rotifers and cladocerans usually reproduce partheno-genetically, with male individuals rare and the population consisting almost entirely of cloned females during periods favourable to growth. Thus egg production is potentially very high. Sexual reproduction is confined to adverse conditions such as low food or low temperature and involves resting, fertilised, eggs. Calanoids and cyclopoids only reproduce sexually, females carrying external egg sacs and as a consequence population growth in these taxa is slower. Cyclopoids possess the ability to undergo resting-stages, known as diapause, usually in the fourth juvenile instar, for prolonged periods during low temperature or food conditions. This provides an alternative method of rapid population growth because diapause is often terminated rapidly by an environmental stimulus (usually temperature) and large numbers of juveniles then synchronously develop into reproducing adults.

The marine environment differs mainly in having few rotifers or cladocera, more extensive representation of protozoa groups, and widespread occurrence of planktonic larvae of sedentary forms in taxa ranging from molluscs and malacostracan crustacea to chordata.

Zooplankton do not depend directly for sustenance upon nutrients, except for microflagellates which possess the facility of both autotrophic and heterotrophic nutrition and may uptake dissolved organic molecules directly (Bird and Kalff, 1986; Stockner, 1988). The majority then are all affected indirectly by nutrient control of the quality and quantity of their algal, bacterial or detrital food. To a lesser extent they may be influenced by the physicochemical conditions of their aquatic environment altered as a result of algal metabolism, such as pH and oxygen concentrations (Smyly, 1978). Species succession, and population growth are thus regulated by the pattern of phytoplankton changes discussed above within the range of a species' occurrence, which is controlled by gross physical features such as temperature range and water renewal time.

The effects of differing food supplies on zooplankton have been extensively investigated. The ecological framework for understanding these effects is the same as that for understanding the effects of nutrients upon algae. This is that a limiting resource (usually food at a defined temperature) determines the population growth rate, r (equivalent to μ for algae), until the carrying capacity, K, is reached, and that different species compete with each other for resources with each species having its own 'niche' described by its suite of resource needs. Species are distributed by their growth rates and physiologies along a continuum between 'r' and 'K' selection, with different strategies succeeding under different conditions of food and physico-chemical conditions. Another way of conceptualising the same strategies was developed by Romanovsky from earlier botanical workers in the USSR and UK (Romanovsky, 1985). This envisages a three-dimensional continuum between three alternatives – 'explerent' strategists (derived from the latin explere, 'to fill') with most rapid growth rates;

'competitive' strategists which maximise resource capture; and 'patient' strategists which are resistant to stressed conditions such as low food supply (Fig. 4.22). The efficiency with which each individual of a species can utilise available resources is determined by the efficiencies of each step in the process of converting food into somatic or reproductive growth. These are feeding (filtering), ingestion, assimilation and respiration.

Food quantity parameters of importance in understanding the food control of growth and competition are the 'threshold concentration', at which population growth rate is zero (Lampert and Schober, 1980) and the 'incipient limiting concentration' of food, at which intake is maximal (Porter *et al.*, 1982), limited by structure of the food-gathering apparatus rather than food supply (Romanovsky, 1985). These differ for each species and size classes within species. Superimposed upon these are food quality parameters determined by the particle size each species can handle (Geller and Müller, 1981), which depends upon individual body size (Burns, 1968), and the food nutritive value (Duncan, 1985). Rates of growth of populations and the outcome of competitive interactions in lakes can be modelled and explained when these parameters are known (Geller, 1985; Kerfoot *et al.*, 1985).

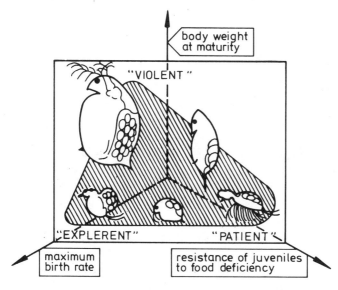

Figure 4.22 The three-dimensional representation of life history strategies in zooplankton. Modified from Romanovsky (1985), with permission.

4.4.2 Competitive displacement of species

At the gross taxonomic level, comparison of species of cladocera, calanoidea and cyclopoidea has shown the latter to be 'macro-filterers', feeding more efficiently on particles between 10 and 100 μm in length, whereas the former are 'microfilterers', operating more efficiently at 5–20 μm. The length of algal particles is not a good indicator of their suitability as food however; particle volume is a better descriptor. This suggests that cyclopoids are more efficient feeders up to 1000 μm^3, calanoids to 100 μm^3 and cladocera around 10 μm^3 (Horn, 1985a,b). Individual species differences within this range, however, are considerable. Geller and Muller (1981) made an electron microscopical study of the mesh sizes of the filtering apparatus of 11 species of cladocera. They showed differences between species which enabled them to classify two with large mesh sizes as 'macrofiltrators' (*Holopedium gibberum* and *Sida crystallina*), four with intermediate mesh sizes as 'low-efficiency bacteria feeders' and five with small mesh sizes as 'high-efficiency bacteria feeders'. They also showed increasing mesh size with growth in body size within species. There is conflicting evidence about whether zooplankton species feed selectively or not; Horn found no selection but Lampert and Muck (1985) found that the calanoid *Eudiaptomus* fed more selectively than the cladoceran *Daphnia*. At high food densities, the latter genus may be able to exert a negative selection by closing the carapace gap to exclude large net-algal species (Gliwicz, 1978), many of which are unpalatable (such as filamentous cyanobacteria). The more important differences between zooplankton species lie in the efficiency with which they feed at certain particle concentrations and size ranges, and in their responses to food limitation.

This combination of morphological and physiological differences in zooplankton species and age classes means that the changing conditions of food particle size and abundance in lakes cause competitive displacement of one species by another. Initially this was simply expressed in the 'size-efficiency hypothesis' (Brooks and Dodson, 1965), part of which stated that larger-bodied individuals were more efficient filter feeders than smaller bodied ones and in the absence of other factors would out-compete them in the plankton. The actual situation is more complex than this because, for example, the juveniles of large-bodied species are highly vulnerable to food limitation whereas adults of small-bodied forms can withstand food depletion by minimising growth and egg production (Romanovsky, 1985). Moreover the specific ingestion rate of food (mg food ingested : mg zooplankton biomass) is higher for smaller zooplankton (Benndorf and Horn, 1985) so that under certain conditions of small-particle dominated food, small-bodied forms can dominate a community.

4.4.3 Effects of predation upon species changes

The effects of food supply on zooplankton species composition can only be considered alone in laboratory experiments; in lakes the effects are almost inevitably compounded with the effects of selective predation. This is more often a size-dependant factor (Brooks and Dodson, 1965; Greene, 1983), so that in lakes with planktivorous fish large-bodied zooplankton are absent or rare, with a consequential greater density of phytoplankton (Timms and Moss, 1984) (Fig. 4.23). Invertebrate predation from large *Cyclops*,

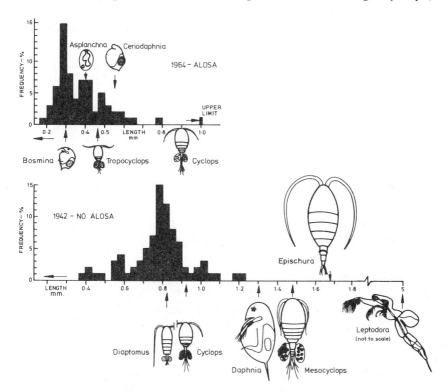

Figure 4.23 The effects of predation and competition upon the dominance and size distribution of zooplankton in lakes, shown by a comparison of a lake with and without a zooplanktivorous fish, *Alosa*. Modified from Brooks and Dodson (1965), with permission.

Leptodora or larvae of the insect *Chaoborus*, may also have an impact on smaller-bodied forms in some lakes (Jamieson, 1980; Pastorak, 1980), particularly in nutrient-rich situations where high zooplankton densities may enhance predator juvenile survival rates in spring leading to high community impact when these predators reach advanced instars in summer (Hall, 1962; Neill and Peacock, 1980).

The combined impact of fish predation and food limitation may explain the disappearence of larger cladocerans in both oligotrophic and eutrophic temperate lakes and in tropical reservoirs (Gliwicz, 1985). In oligotrophic lakes, high water transparency combined with the absence of a littoral refuge could make fish predation pressure high. Food limitation, particularly during the early part of the year when ephippial eggs might hatch, also seems necessary to explain the complete absence of large cladocerans from ultra-oligotrophic lakes. In eutrophic lakes, increase in algal biomass does not necessarily mean an increase in available food because of the increasing dominance of unpalatable algae; consequently, food limitation of juveniles combined with predation of adults can again be sufficient to eliminate species. In tropical reservoirs, which often contain a pelagic zooplanktivorous sardine species, predation is avoided by diurnal migration patterns but the metabolic costs of this activity may be unsustainable during periods of low food availability, leading to the disappearance of large-bodied species.

4.4.4 Seasonal succession of zooplankton and effects of eutrophication

Oligotrophic and mesotrophic lakes are often dominated by calanoid copepods whose population shows a gradual increase through spring to a summer peak followed by a gradual decline, following the similar pattern of algal biomass. Calanoids are believed to dominate because they are more efficient feeders on the larger algal species characteristic of these lakes, and feed with higher ingestion efficiency at low food density (McNaught, 1975). As food density increases in eutrophic lakes larger-bodied clado-cerans may become more abundant if fish predation is not too high. Cladocerans have higher population growth rates than calanoids so peak earlier in spring if food supplies are adequate. George *et al.* (1990) have shown that the cladoceran *Daphnia* became more abundant, and the calanoid *Eudiaptomus* less abundant as eutro-phic species of algae increased in Estwaite Water in the English Lake District. The population cycles of *Daphnia* during the warmer months of the year were controlled by the relative proportions of edible and inedible algae. Cladocerans are more efficient filter feeders at moderate–high concentrations of food because there is more food ingested for the same amount of energy expended compared with low-food conditions (Richman and Dodson, 1983). Predation may change the balance in favour of smaller-bodied cladocera such as *Bosmina* instead of *Daphnia* (McNaught, 1975). The seasonal patterns of rise and fall of populations may become more erratic as short-term reduction in food abundance causes rapid mortality of juveniles but equally rapid parthenogenetic egg production by surviving adult females when food increases. With further increases in nutrient status bacterial abundance also increases and high efficiency bacterial feeders, such as the large-bodied *Daphnia magna* (Harper, 1986) or the small-bodied *Ceriodaphnia* become dominant, depending upon predation pressure. However, high concentrations of cyanobacteria are less nutritious (Porter and Orcutt, 1980) and cause blockage of the filter-feeding apparatus in large-bodied forms, making energy expenditure exceed intake. Under such conditions, usually in hypertrophic lakes, the more selective feeding of calanoids may again give them a competitive advantage (Richman and Dodson, 1983). For the same reasons, the seizing techniques of herbivorous cyclopoids may also enable them to increase in abundance. Small-bodied cladocera, particularly the normally littoral chydorid *Chydorus sphaericus*, may become temporarily abundant at the height of cyanobacterial population growth and decay because of their ability to filter efficiently the fine particles of bacteria and cyanobacterial cells associated with the bloom without mesh-clogging by the large colonies.

Seasonal succession of zooplankton grazing types in eutrophic lakes may mirror the pattern of dominance between lakes with change in trophic state (Gliwicz, 1977), with macrofiltrators dominant during low food concentrations (oligotrophic lakes and the beginning of the year), giving way to low-efficiency algal and bacterial filter-feeders in spring blooms of mainly edible algae and high-efficiency bacterial feeders in summer blue-green blooms. Overall biomass of zooplankton increases with trophic state over a wide range of temperate and tropical lakes (Patalas, 1972; McCauley and Kalff 1981; Bays and Crisman, 1983), and with an increasing dominance of small-bodied forms.

4.5 PRODUCTION AND SPECIES CHANGES OF ZOOBENTHOS

4.5.1 Introduction

The bottom fauna, or zoobenthos, are animals inhabiting the substrates available in lakes and rivers. Two zones are usually recognised: the first from immediately below the water surface at the edges to the depth to which light penetrates to sustain rooted plant growth (the littoral); and the deep water sediments (the profundal). There are a variety of substrates available for animal colonisation; these are grouped into two types of habitat: surfaces such as the plants themselves, stones to which sessile animals may attach themselves, and the sediments into which many animals burrow. Some mobile animals, such as gastropod snails, may move from one type to another but generally the adaptations of species confine them to one type. Lakes, rivers and marine environments share many characteristics in common, and excellent comparative reviews of the benthos are provided by Mann (1980) and Lopez (1988).

The bottom of a lake or river is subject to a variety of physico-chemical conditions which determine species' occurrence. Chief of these is turbulence, which controls the particle-size distribution of sediments. Generally fine particles are removed from the wave-influenced upper parts of the littoral zone and deposited in deeper water, although the presence of macrophyte beds (themselves influenced by wave action) may modify this. Hence burrowing and tube-dwelling species tend to be more common in the profundal (and pools of rivers) whilst surface-attached forms are more common in the littoral. Temperature varies between the two zones, with a gradient between more extreme fluctuations determined by the atmosphere experienced in the littoral and more constant conditions determined by the water mass in the profundal. Oxygen and pH levels are the more important chemical conditions, with the effects of stratification determining the biochemical processes of the sediments and oxygen concentrations in particular affecting the ability of essentially immobile individuals of a species to survive. In the context of oxygen concentrations, it may be difficult to separate the effects of eutrophication and organic pollution if both occur together (as for example with influx of sewage effluent to a lake) because both increase organic matter and bacterial decay, decrease oxygen concentration and at the same time increase the rain of organic detritus to the benthos, which is the primary source of food.

Many benthic animals exhibit adaptations to low oxygen conditions, such as haemoglobin or other oxygen-binding blood pigments combined with distinct ventilatory movements in midge larvae of the genus *Chironomus*.

Biological factors affecting benthic species are primarily those which determine food supply, and since few benthic herbivorous species can actively seek food, this is largely detrital. In the littoral zone detrital food may be derived from the decay of macrophytes, but here epiphytic algal growth on plant and stone surfaces may support grazers such as gastropod molluscs and some insect larvae. In the profundal the major food supply is fine particulate material from the epilimnion together with a proportion of macrophyte detritus exported from the littoral. Competition for the available food, and predation effects upon herbivores, are additional factors influencing the structure of benthic communities but there is far less information upon the precise nature of these factors than there is for zooplankton and algal communities.

The nature (shape and size) of the food supply is the primary determinant of the mechanisms by which invertebrates gather their food. The original source is almost entirely as a rain of fine particles so many animals are filter-feeders, collecting particles before they reach the bottom. Bivalve molluscs establish feeding currents with protruding siphons; many chironomid larvae create feeding (and respiratory) currents by rhymical body movements through mud tubes. Other organisms, such as oligochaete worms, ingest the mud particles and their associated microbial colonists. Some species collect from the surface, such as amphipod and isopod crustacea. The faeces of all organisms are re-deposited on the sediment surface and form part of the food supply.

The overall biomass of benthic invertebrates depends upon the quantity of food supplied. Since all benthic organisms must wait for their food to arrive, the extent of decay it has undergone before it arrives is important; the longer it takes, the more microbial mineralisation will have taken place in the water column. Hence there is generally a connection between depth of water body and benthic biomass, with deep lakes (which may also have lower primary productivity) always having lower benthic biomasses (Deevey, 1941). Lakes generally have higher overall benthic biomass than rivers, because rivers are subject to greater physical disturbances and the constant unidirectional 'spiralling' of organic matter towards their mouths.

4.5.2 Taxonomic groups and life history patterns

The main taxonomic groups in the macrobenthos are insect larvae or nymphs, with terrestrial adult stages; crustacea; worms and molluscs. The microbenthos is largely composed of crustacea (similar taxa but different species to the zooplankton), rotifers and protozoa. There are also a variety of free-swimming forms, chiefly crustacea and rotifers similar to planktonic forms, associated with (or temporarily attached to) surfaces from which they sweep food by filter-feeding mechanisms. The littoral zone supports the largest number of species, particularly in macrophyte beds where the combined habitats of plant and sediment offer most ecological niches, with up to 300 species in Lake Esrom for example (Jónasson, 1978), dominated by molluscs and a variety of insect larvae. Generally the profundal is much less diverse than the littoral, falling to less than 20 species in Lake Esrom, with dominance of chironomid midge larvae and oligochaete worms. The profundal may have the highest number of individuals, composed of small, slow-growing forms, but biomass is often highest in the intermediate zone between the littoral and profundal, corresponding to the zone within which the thermocline occurs during stratified periods. Here the supply of 'fresh' detritus from the littoral zone is expoited by larger-bodied detrivores such as the crustacea *Asellus* and bivalve mollusc *Dreissena* (Fig. 4.24).

Most macrobenthic animals have distinct generations through the season (punctuated by adult emergence in insects) with life spans ranging from a few months to one or two years; microbenthic animals are more similar to planktonic forms with continuous reproduction and the number of generations determined by temperature. The length of a generation may be determined by food supply as well as temperature, with peaks of growth in profundal invertebrate populations corresponding to the period of sedimentation of declining phytoplankton populations (Jewson *et al.*, 1981; Rasmussen, 1984; Jónasson, 1978) (Fig. 4.25). Growth may cease in low oxygen conditions if oxygen supply is only adequate to sustain ventilatory but not food-gathering energy expenditure (Jónasson and Kristiansen, 1967) and some species may be able to tolerate anoxia for long periods. The oligochaete *Tubifex tubifex* was able to survive, grow and reproduce under continuous anoxia for 10 months (Famme and Knudsen, 1985).

Measurement of the production of species with distinct cohorts, such as many benthic invertebrates (and fish, see below) is made easier by the ability to recognise juvenile stages and follow their development. This is generally referred to as the Allen Curve technique, after the New Zealand fisheries ecologist who first developed it (Allen, 1951). Numbers of individuals, which progressively decline after hatching through predation and other causes of death, are plotted against mean weight, which increases through individual growth, and the area under the resultant curve between two points is net production. Jónasson compared the production of the chironomid midge *Chironomus anthracinus* with the oligochaete worm *Potamothrix hammoniensis* (the two dominant species) in Lake Esrom, Denmark, using this method (Jónasson, 1975). He found that *C. anthracinus*, which is a surface feeder, had a faster growth rate, shorter life cycle (1–2 years) and higher production than *P. hammoniensis*, a sub-surface feeder with a 4–6 year life cycle (Fig. 4.26).

4.5.3 Effects of eutrophication

Eutrophication may have different effects upon the littoral and profundal zones of lakes, with changes in oxygen depletion affecting the littoral zones more than the profundal. Littoral zone changes may mirror the changes in rivers subjected to organic pollution, with progressive decline of oxygen-sensitive species as more of the hypolimnion becomes deoxygenated (and thus a larger area of littoral zone comes under the influence of low oxygen) or if the whole lake experiences reduced oxygen concentrations at overturn. Harper (1986) showed a decline in invertebrate taxa in a mesotrophic–hypertrophic shallow lake series in Scotland from 61 to 33, with progressive disappearance of stonefly, mayfly nymphs and caddisfly larvae from the littoral zones. Littoral biomass showed little differences between the lakes, but profundal biomass increased by about 50% in the eutrophic and hypertrophic lakes, with a progressive decline in the proportions of chironomid larvae to oligochaetes.

Macan has shown that changes in species composition in the littoral zone may occur through changes in predation intensity as an indirect effect of enrichment (Macan, 1980). Oligotrophic lakes are dominated by the large stonefly predator *Diura*, which suppresses the populations of potentially competing flatworm predators. With an increase in production of periphytic algae and detritus, detritivores such as *Asellus* increase in density and 'escape' from predation by rapid growth to large-sized individuals less easily

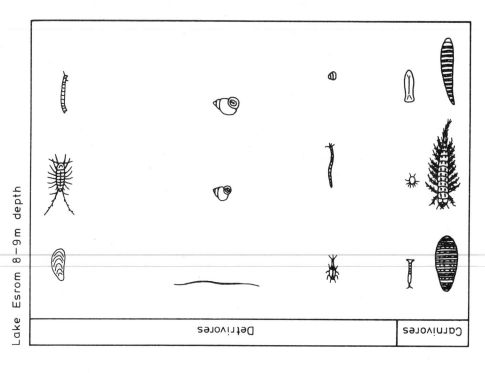

Lake Esrom 8–9m depth

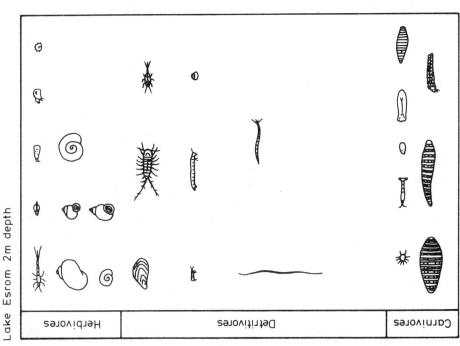

Lake Esrom 2m depth

Lake Esrom 20 m depth

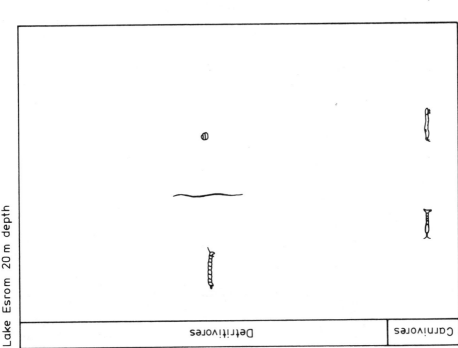

Detritivores

Carnivores

Figure 4.24 Typical benthic invertebrates of the littoral and benthic zones of Lake Esrom, Denmark, showing decrease in species richness with depth. At 2 m depth, the taxa illustrated are from top to bottom and left to right: *Centroptilum, Micronecta, Oxyethira, Eurycercus, Armiger, Lymnea, Valvata, Planorbis, Gyraulus, Bythinia, Dreissena, Asellus, Caenis, Cathocamptus,* Chironomidae, *Pisidium,* Oligochaeta, Ceratopogonidae, Tanypodinae, *Candona, Polycelis, Helobdella, Glossiphonia, Erpobdella,* Leptoceridae. The additional taxa at 8-9 m is *Sialis* and at 20 m *Chaoborus,* with the chironomid and oligochaete at 20 m being *Chironomus anthracinus* and *Potamothrix hammoniensis* respectively. Modified from Jonasson (1978), with permission. (The original shows relative abundances which have been omitted here for simplicity.)

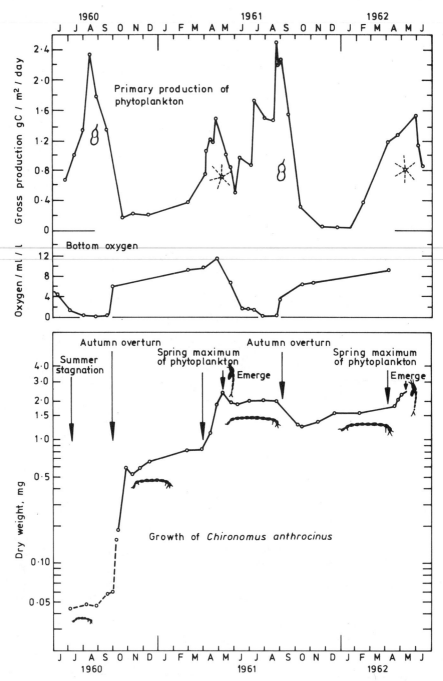

Figure 4.25 Growth of *Chironomus anthracinus* in relation to phytoplankton production and oxygen concentration in Lake Esrom. Modified from Jónasson (1978), with permission.

caught. Stoneflies decline because their eggs are scattered hap-hazardly, so are at increasing risk of consumption by detritivores. Flatworm predators increase because they have continuous growth throughout the summer compared to the slow growth and single generation of *Diura*, so the latter is out-competed as well as suffering from direct interference.

In a review of the effects of eutrophication on the profundal fauna of deep lakes, Sæther was able to draw together a number of general effects (Sæther, 1980). The earliest effect is an increase in abundance of all groups, particularly the oligochaetes, bivalve *Pisidium* and larger crustacea, but without any changes in species composition. The proportions of oligochaetes to other taxa in the community progressively increases. There are then changes in the relative abundance of dominant chironomid species, with the disappearance of some species which are oxygen-sensitive and thus characteristic only of oligotrophic lakes. The family chironomidae has so many species (> 400) that it has been possible to draw up an extensive classification of 15 community types which indicate progressive change from extreme oligotrophic to hypertrophic lakes in the Old World and their taxonomic equivalents in the New (Sæther, 1975, 1979) (see Chapter 6). The changes are proportional to increases in both total phosphorus loading/mean depth and chlorophyll 'a'/mean depth. Changes in chironomid communities occur most rapidly in the initial stages of transformation between oligotrophy–mesotrophy, but lag behind the changes in nutrient and primary producer relationships because of the modifying effect of depth. Thus Lake Mjøsa, Norway's largest lake, is considered eutrophic on phosphorus loading and chlorophyll characteristics (Holtan, 1980) but still retains a strongly oligotrophic benthic chironomid community.

There are limited examples of the effects of reversal of nutrient-rich conditions upon benthic macroinvertebrates, but in those that exist, reversal of the trends outlined above has been observed. Thus in the Bay of Quinte, a long thin arm of Lake Ontario, reduction of phosphorus loadings after 1977 resulted in a decline in numbers and biomass of oligochaete worms, sphaeriid molluscs, isopod crustacea and some chironomids. Dominance in both chironomid and oligochaete communities changed towards species less tolerant of eutrophic conditions (Johnson and McNeil, 1986).

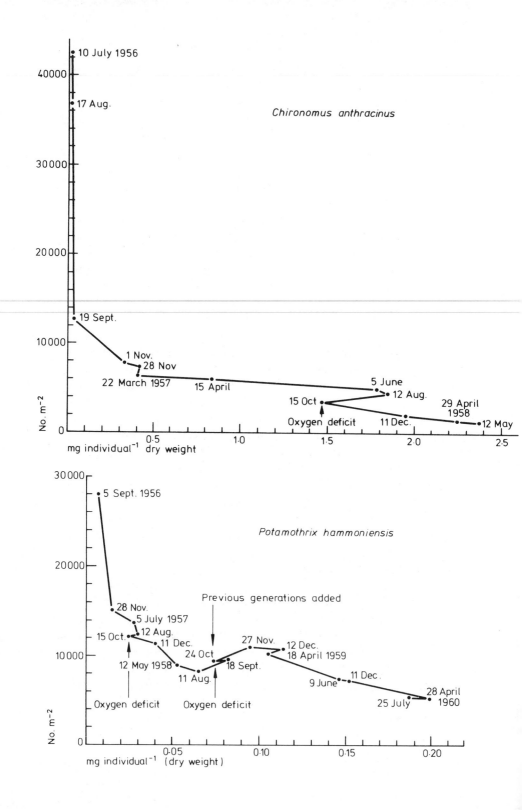

4.6 EFFECTS OF EUTROPHICATION ON FISH AND OTHER VERTEBRATES

4.6.1 Introduction

Fish are generally in the upper levels of the aquatic food web and are thus affected directly or indirectly by several levels below them. The fish community in a lake may, however, include representatives of all trophic levels including herbivores and detritivores. Many species show sequential changes from herbivory to predation through their life cycle. Fish are mobile, and thus capable of responding to altered biological or physical conditions by migration. They have long life-spans, ranging from a few years to several decades for predators such as pike and salmonids. Fish are also the only taxa in the aquatic ecosystem directly exploited by man. Their history of exploitation in lakes is often one of changes caused through deliberate or accidental management practices. Over-exploitation may alter the competitive balance between species and widespread introductions of non-native fish have often had similar competitive-mediated changes. Introductions have also influenced parasite and disease susceptibility of native fish stocks. The history of the fisheries of the Laurentian Great Lakes in North America provides a good example of the complex effects of overfishing, shoreline development, effects of introduced fish and parasite species and nutrient enrichment (Goldman and Horne, 1983); consequently, effects of eutrophication are often masked by other biological changes. Their position in the upper levels of food webs also means that fish may exert a marked influence on the structure of the levels below them.

Fish populations are relatively hard to sample accurately compared to the taxonomic groups discussed previously in this chapter, because they are mobile. Capture methods range from nets, and trawls for samples and electric fishing and poisoning for total removal. Their biology, however, is relatively easy to study for several reasons. They are large, have distinct age classes (often year-classes in temperate regions), individuals can be marked and recaptured, and growth is temperature-dependent, hence often pulsed. This can be followed in the examination of fish scales which

Figure 4.26 Production of *Chironomus anthracinus* and *Potamothrix hammoniensis* in Lake Esrom, estimated by the Allen Curve method. Modified from Jónasson (1975), with permission.

record periods of growth and quiescence just like tree rings. Paradoxically, the difficulties of sampling have often led to their neglect in limnological studies whilst their importance to man has led to a wealth of information about fisheries and their management. The two strands of aquatic research are rapidly coming together now, not least because of the increasing interest in 'biomanipulation' of eutrophic lakes – the management of fish stock to control the lower trophic levels and alleviate the worst effects of enrichment.

Good introductions to fish biology are provided in Goldman and Horne (1983) and Moss (1988); there is more detailed treatment in Weatherley (1972), Gerking (1978) and Pitcher and Hart (1985) and methodology in Bagenal (1978). There is an almost separate development of the science of deliberate eutrophication of waters (often small ponds) to provide maximum fish production for human utilisation, aquaculture, which has been widely practised in some parts of the world for centuries.

Fish occupy and feed in all habitats of the aquatic environments, with many species moving between two or more on a daily or seasonal basis in relation to both feeding and reproductive patterns. In extremes are species such as eels or salmon moving between marine and freshwater systems. Fish are less cosmopolitan however; more than any other taxa, they show distinct regional differences in occurrence, with some families exclusively tropical and many more species in the tropics with endemics in large lakes and river systems.

Fish may be conveniently classified as pelagic (or open water) species, where they are either planktivorous or piscivorous feeders; littoral, where they may be herbivores, detritivores, or carnivores; and benthic, where they are detritivores or carnivores. Most temperate fish breed in the littoral zone, attaching their eggs to surfaces of stones or plants (cyprinids) or scattering them in gravel beds (salmonids). Tropical fish have many different spawning strategies, ranging from the production of enormous numbers of planktonic eggs (Nile Perch) to mouthbrooding eggs and newly-hatched young (cichlids).

The recruitment success of newly hatched fish often depends upon the short-term combination of favourable environmental and biological conditions, and the effects of these conditions can be followed in that age class of fish as it develops throughout the life span of the species. There is often little relationship between the number of eggs laid and the number of fish recruited to the stock at the end of the first two years of growth (Craig and Kipling, 1983).

Thus in Windermere, England, 300-fold fluctuations in recruitment of perch have been observed over a 41-year study period during which biomass of parental stock varied 6-fold and number of eggs laid only 2-fold. This recruitment success was strongly correlated with temperature in the first year of life. Pike recruitment varied only 7-fold and was less dependent upon temperature, probably buffered by the variety of food supply since pike switch from planktivorous to piscivorous food within a few weeks of hatching.

4.6.2 Effects of eutrophication

Direct effects of oxygen concentration changes

Fish families may be arranged in a series of decreasing susceptibilites to oxygen reduction, with salmonids and coregonids dependent upon high oxygen concentrations, down to cyprinids, some of which tolerate very low concentrations (Doudoroff, 1970; Willemsen, 1980). The precise tolerance, however, depends upon species and age of the fish, and external factors such as temperature, time of exposure and previous history. Eutrophication causes an increase in plant production in lakes; this may result in supersaturation of lake waters during daytime photosynthesis and depletion by night-time respiration. Day-time photosynthesis may also elevate pH to lethal values affecting gill function. The ultimate consequence of plant production is an increase of detritus and hence bacterial oxygen demand, which itself may deplete oxygen concentrations, particularly at high temperatures. Adult and juvenile fish can avoid low oxygen conditions by movement out of affected areas, for example out of deeper waters or weed zones, but they may then encounter other unfavourable conditions. In deep lakes subject to eutrophication cold water fishes (coregonids and salmonids) may be unable to tolerate high temperatures in the epilimnion when they are driven there by progressive deoxygenation of the hypolimnion (Colby *et al.*, 1972). Their eggs in littoral gravels are more sensitive to short-term oxygen depletion and to the oxygen demand of increased detritus, so initial effects of eutrophication may be reduction in spawning and recruitment.

Several examples exist of sudden and complete fish kills in nutrient-rich lakes as a consequence of algal-mediated deoxygenation. This occurs in shallow lakes at high temperatures, where the nightime oxygen demand of dense phytoplankton blooms, if mixed throughout the lake, depletes the entire water column. It may also occur under ice during winter when sunlight is high enough to stimulate an increase in algal biomass but the ice prevents replenishment of oxygen used up by respiratory processes (Schindler, 1987).

Biomass and taxonomic changes

The overall population sizes and biomass of fish increase with enhanced nutrient supply. In some studies this has been demonstrated clearly with little evidence of subsequent competitive interaction between species. In Lake Memphremagog, for example, a long thin lake on the US–Canadian border with point source nutrient inputs at its southern end, abundance and biomass of littoral zone fishes showed similar gradients from south to north as other indices of productivity in nutrients and plankton. Biomass was approximately three times greater in the south, with less inter-specific competition (Gascon and Leggett, 1977). In larger lakes, a similar overall increase may mask different changes in individual species: in Lake Constance on the Swiss–Austrian–German border, progressive eutrophication resulted in a five-fold increases in fish catches with faster growth rates in planktivorous cyprinids but restricted growth of coregonids in the most eutrophic basin and a decline in their overall proportion of the fish community (Nümann, 1973). Overviews of the effects of eutrophication on fishery yields are provided by Hartmann (1977) for European, Colby *et al.* (1972) for North American lakes and Leach *et al.* (1977) for percid species on both continents (Figs. 4.27–4.29).

Ecosystem interactions with eutrophication

Direct effects of eutrophication occur on the magnitude and balance of primary production. In many cases this decreases the extent, biomass and production of rooted aquatic plants and it is the state of the rooted plant community which directly influences the fish community. In European lakes, the semi-natural, more oligotrophic state in both deep and shallow lakes is characterised by a zonation of littoral plant communities from emergent through floating-leaved to submerged species with increasing depth, the extent of depth colonisation of submerged species dependent upon light penetration. The fish communities (Fig. 4.30) are dominated by the piscivorous predator pike (*Esox lucius* L.), which requires shelter to 'hide-and-wait' for prey, piscivorous/invertebrate-feeding perch (*Perca fluviatilis* L.) and invertebrate-feeding tench (*Tinca tinca* L.). Weed fringe/open water zones are characterised by invertebrate-

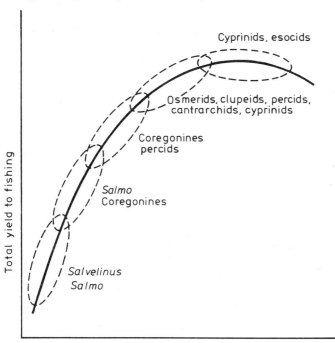

A Morphoedaphic Index

Figure 4.27 Trends in fish yields and taxonomic composition with eutrophication. Total yield and its taxonomic composition in North American lakes. Modified from Colby *et al.* (1972), with permission.

and plant-feeding rudd (*Scardinius eryopthalmus* L.), roach (*Rutilus rutilus* L.) and more open water by bottom feeding species such as eel (*Anguilla anguilla* L.) and ruffe (*Gymnocephalus cernuus* L.). Open water of colder, deeper lakes also contain salmonids such as *Salmo trutta* L. and coregonids such as *Coregonus albula* L.; species with catholic feeding habits amongst both benthos and plankton. Species which prefer more open, turbid environments (naturally the lower reaches of rivers) are absent or rare in clear-water, weed-fringed lakes. Such open environments are dominated by pisci-vorous pike-perch (*Stizostedion lucioperca* L.), benthivorous bream (*Abramis brama* L.) and carp (*Cyprinus carpio* L.) and plankti-vorous smelt (*Osmerus eperlanus* L.) (Grimm, 1989).

In lakes with aquatic vegetation-dominated littoral zones preda-tion pressure from pike maintains low biomass of the cyprinid fish and hence low predation pressure on zooplankton; the latter exert a strong grazing pressure on phytoplankton algae leading to clear

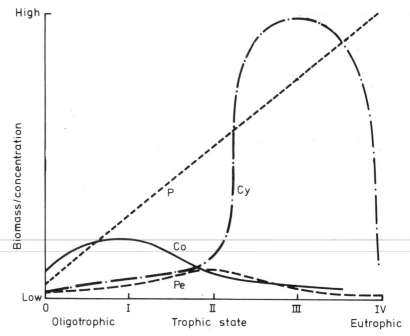

Figure 4.28 Yield of fish families in European lakes with increasing phosphorus concentration. Modified from Hartmann (1977), with permission. M = macrophytes, P = phosphorous, B = total biomass, Co = coregonids, Pe = percids and Cy = cyprinids.

water which allows suitable light penetration for the maintenance of both rooted plant growth and high visibility for pike feeding. Progressive reduction in vegetation has two effects. Firstly, by the reduction of its carrying capacity for pike, predation pressure on planktivores is reduced. Most species of cyprinid have planktivorous fry so the depletion of grazing zooplankton can be rapid during the spring, resulting in increased algal biomass and decreased light penetration. Decreased light has a negative effect on rooted vegetation growth and the detection of prey by pike. Secondly, reduction in the area colonised by vegetation increases the area of open mud bottom in the lake, increasing the feeding area for benthivorous bream and related species. These too have a negative effect on transparency, by the resuspension of sediment caused by their mode of feeding (Lammens, 1989). This is both direct through the effect of sediment particles in the water column and indirect through the recycling of sediment nutrients which are released for further phytoplankton growth. Roach and white bream also feed on benthic invertebrates, but on smaller individuals less deep in the

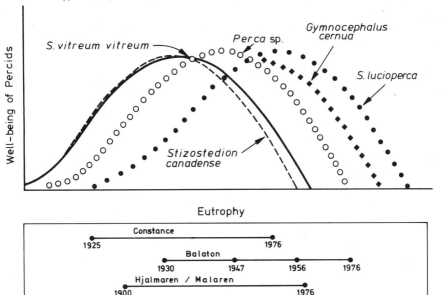

Figure 4.29 Change in composition of percids with eutrophication in lakes in Europe and North America. Modified from Leach *et al.* (1977), with permission.

sediments; moreover bream are less vulnerable than the other two species to predation because they grow more rapidly. In turbid waters the only piscivore is pike-perch but its impact is confined to smaller-sized individuals under 25 cm. Lakes where piscivore predation is low, such as those in Britain where pike-perch are naturally absent or where human fishing pressure is intense, may be dominated by bream and large populations of often small-sized cyprinids such as roach whose planktivore predation is consequently high, resulting in phytoplankton growth characterised by high biomass and persistence of cyanobacteria.

Studies in individual lakes have demonstrated the extent to which the general picture given above is more complex, and modified by particular conditions. North American lakes generally support a greater range of fish families but the overall pattern of piscivore-driven changes as a consequence of eutrophication is similar (Leach

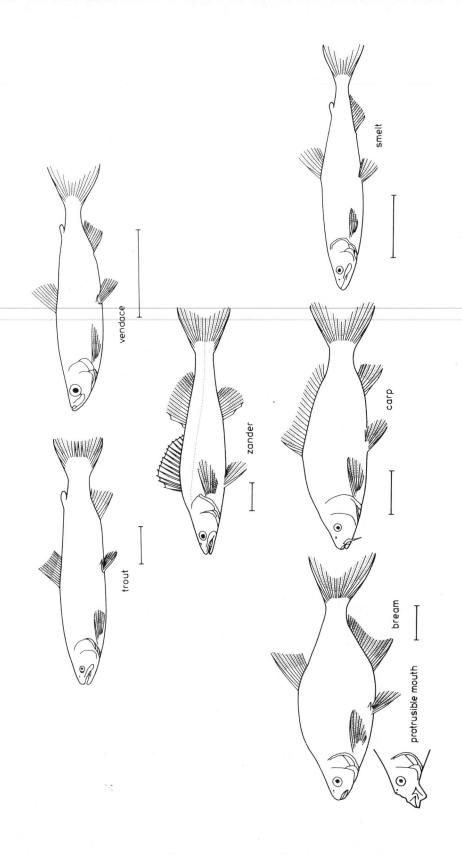

smelt

vendace

zander

carp

trout

bream

protrusible mouth

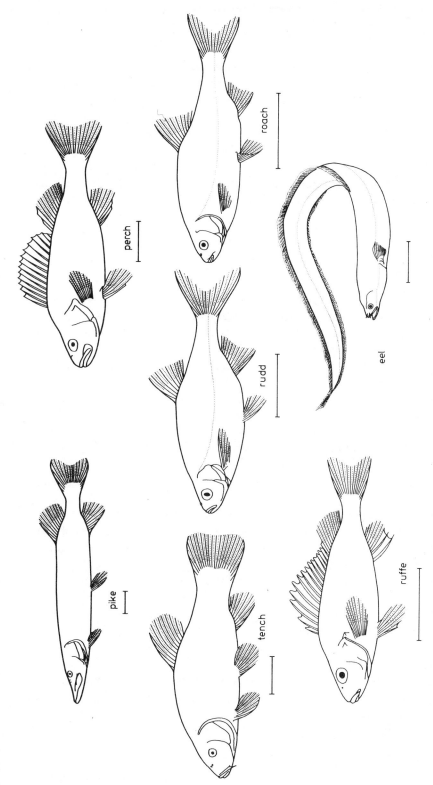

Figure 4.30 Examples of the fish of European lakes of differing trophic state (size bars = 10cm). Modified from Wheeler (1978), with permission.

et al., 1977; Hurley, 1986). At more detailed levels of feeding interactions there may be alternative pathways; for example either roach fry or carnivorous invertebrates such as *Cyclops* or *Mysis* can maintain predation pressure on herbivorous cladocerans and thus release phytoplankton from grazing suppression (Braband *et al.*, 1986).

On a macro-scale, fishery yields, particularly from lakes where the entire fish stock is exploited, such as lakes in eastern Europe, provide a larger geographical perspective of the effects of eutrophication. The trends are similar: decline in salmonid fish and other piscivores and increase in coarse fish of lower economic value (Leopold *et al.*, 1986).

4.6.3 Other vertebrates

Higher vertebrates, particularly aquatic birds, are relatively independent of any one aquatic environment and can move away from less favourable conditions or accumulate in favoured ones. Nevertheless some general effects of eutrophication on birds and mammals can be discerned. Increases in biomass of aquatic vegetation with mild eutrophication have been linked with increasing numbers of herbivorous species such as moorhen (*Gallinula chloropus* L.) and whooper swan (*Cignus cignus* L.) in Finland (Haapanen, 1978; Pullianen, 1981). Some piscivorous birds have also shown increases, such as the great-crested grebe (*Podiceps cristatus* L.) in Denmark and Western Germany (Asbirk and Dybbro, 1978; Flade, 1979). Similar increases of grebe and swan have been linked with moderate eutrophication in southern Germany, albeit with a reduced overall species diversity of waterfowl (Utschick, 1976). However, enrichment may not be the sole cause because numbers may also have increased as a result of an increase in area of suitable new habitats such as gravel-working lakes, and densities on eutrophic lakes could be a consequence of dispersal rather than a reflection of conditions in the lakes themselves. Kerekes has suggested that the density of breeding common loons (great northern diver; *Gavia immer* Brünn.) in parts of Canada is a reflection of trophic state as indicated by fish production (Kerekes, 1991).

More advanced stages of eutrophication in lakes, where aquatic vegetation changes have been recorded, are detrimental to birds. Pochard (*Aythya ferina* L.) have declined on Lake Constance as their main food supply, the macroalga *Chara*, has declined with eutrophication. General macrophyte decline has been linked with the progressive disappearance of black swan (*Cygnus atratus* L.) and Canada geese (*Branta canadensis* L.) in Ellesmere Lake, New Zealand (Hughes *et al.*, 1978), and with mute swans (*Cygnus olor* Gm.), coot (*Fulica atra* L.), teal (*Anas crecca* L.) and gadwall (*Anas strepera* L.) decline in Loch Leven, Scotland (Allinson and Newton, 1974). In other eutrophic lakes where the water quality effects of eutrophication are clearly documented, the effects on birds are less clear because of the contributory factors of human pressures and wetland or shoreline disturbance. Potential eutrophic effects on mammals are even more likely to be obscured by these influences (Crowder *et al.*, 1986).

In certain circumstances birds may be contributory factors in eutrophication where they breed or roost in high densities. A night-time roost of black-headed gulls (*Larus ridibundus* L.) for example, provided most of the phosphorus input to one of the Norfolk Broads, England, during the 1970s (Leah *et al.*, 1978).

4.7 AQUATIC FOOD-WEB CONSIDERATIONS

An overall understanding of the effects of eutrophication upon the whole aquatic ecosystem is difficult to achieve, for two main reasons. On the one hand, the effort involved in understanding the effects of nutrient enrichment at a single site becomes progressively more expensive with passage up the food web from producers to consumers (Morgan and McLusky, 1974), and the effects of other human influences such as shoreline modification and harvesting or control (e.g. of fish or macrophytes) have also to be taken into consideration. On the other hand, comparative studies of data sets from a large number of ecosystems may show some statistically significant abiotic-productivity relationships but with wide scatter of points due to individual lake characteristics and different methods of data collection and less significance at higher trophic levels (Brylinsky, 1980).

There is a clear trend of increase in phytoplankton biomass with increase in nutrients; usually the best regression is with total phosphorus (McQueen *et al.*, 1986). There is also a significant relationship between phytoplankton production and biomass, and between both biomass and production and photosynthetic efficiency (Brylinsky, 1980). Thus productive, eutrophic lakes generally utilise sunlight more efficiently than oligotrophic lakes. However their turnover time is longer and hence their productivity/biomass (P/B) ratio may be lower because of the greater importance of large, colonial species such as cyanobacteria (Riemann *et al.*, 1986). At the secondary producer level, biomass of zooplankton is positively correlated with phytoplankton chlorophyll (McQueen *et al.*, 1986) and production of both zooplankton and zoobenthos positively correlated with phytoplankton production, with higher turnover times in more productive lakes (Brylinsky, 1980; Riemann *et al.*, 1986). Zoobenthos production is also correlated well with the inverse of mixed depth, indicating that shallow, well mixed lakes are more likely to channel primary production into benthic communities probably because of the shorter retention time of detritus in the water column.

Correlations between zooplankton biomass and planktivorous fish are much weaker, and this has been interpreted as a result of the development of two alternative states in nutrient-enriched rich lakes (Hillbricht-Ilkowska, 1977; Sprules and Knoechel, 1984). In oligotrophic lakes, zooplankton feed directly upon phytoplankton and are large-bodied in form because planktivore predation is kept low by piscivores. Initially, increasing trophy increases biomass at each level. At a certain stage piscivore predation is reduced, possibly associated with decline of macrophyte habitats and planktivore predation then causes a shift to smaller bodied zooplankton which do not graze phytoplankton directly but associated bacteria and detritus. Energy which formerly flowed through planktivorous fish is shifted (either by habit changes or species shifts) more into benthivorous and direct phytoplanktivorous feeding (McQueen *et al.*, 1986; Riemann *et al.*, 1986), but phytoplankton biomass is higher and more phytoplankton production is recycled through bacteria and ultimately sedimented.

These considerations of trophic structure and function form the basis for the practical management of eutrophic lakes. They give two important guidelines for management. The first is that primary production is controlled from the 'bottom' of the food web − by nutrient supply − mainly total phosphorus. The second is that the way in which production is utilised depends upon conditions at the 'top' of the food web − by the effects of fish predation on planktivorous fish and the grazing impact of the latter upon zooplankton. Effective management of eutrophication requires that attention is given to both these controls and not just one of them (Chapter 7).

4.8 WIDER IMPLICATIONS FOR WILDLIFE AND CONSERVATION

The overall value of a lake for conservation may be compatible with its value for other uses, but not always so. Few lake studies have focussed solely upon the conservation value of lakes undergoing eutrophication because the stimulus for such studies has often been measured more directly in economic terms, such as the loss of fisheries, potable water supply or recreational income. As a generalisation, eutrophication decreases conservation value, because it usually results in fewer species of any taxonomic group, together with increased dominance of one or two species. Conservation value is a subjective judgement which is usually applied to species richness and rarity value of larger plants and animals (Usher, 1986); thus, as far as eutrophication is concerned, the effects upon aquatic rooted plants, larger invertebrates such as dragonflies, and birds are of most concern (Morgan, 1970b). Harper (1986) showed how increase in trophic state in a loch series in Scotland was accompanied by decrease in species richness in the algal, macro-phyte and invertebrate communities while studies of single sites which have become increasingly enriched over time have recorded similar declines in species richness (Morgan, 1970a). Distribution of rare species, such as coregonid fish in relict populations from the last ice-age in Britain, are an example of elements of fauna at particular risk from eutrophication but in many other cases the greater conservation concern is over habitat modification, such as wetland drainage (albeit indirectly connected with eutrophication), as in the Norfolk Broadland.

5

The engineering, economic and social effects of eutrophication

5.1 INTRODUCTION

The problems of eutrophication are felt most strongly where human economic and public health interests, rather than merely conservation or aesthetic interests, are affected by its consequences. Thus the earliest recorded manifestations of eutrophication in Europe and North America were in those lakes which were utilised as important sources of potable water and commercial fisheries, such as lakes Zurich and Erie. Over the past two decades in developed areas, particularly in the northern hemisphere, recreational demands have increased so the effects of eutrophication upon such activities as boating and sport fishing may now be measured in economic terms.

5.2 WATER SUPPLY

5.2.1 Historical background

The supply of piped potable water to domestic and industrial premises from centralised sources began around the middle of the last century in Europe and America as a consequence of the need to provide disease-free supplies of water separated from waste disposal. It has accelerated, particularly in rural areas, in the last four decades (Parker and Penning-Rowsell, 1980). This in turn has been followed by the development of mains drainage and sewage treatment to communities formerly served by septic tanks or similar systems. In many cases groundwater supplies can be utilised for potable supply but these are often of finite quantity even in suitable geological areas. River water supplements well-derived supplies, but fluctuates seasonally in quantity as well as quality. Consequently most supply development has involved impoundment of river supplies (Henderson-Sellars, 1979). It has been recognised from the beginnings of water supply planning that impoundment of a surface water source not only produces a more regular supply but also that storage is an effective first treatment process causing sedimentation of bacterial and other particles.

In Britain, impoundment of rivers by the damming of valleys began in the first half of the last century and reached a peak in the second half, serving the conurbations which developed after the Industrial Revolution. The majority were in upland areas which offered higher rainfall, low population density and hence easier aquisition of the land needed together with unpolluted river sources. The largest costs were often associated with the supply of piped water over several hundred kilometers to the supply zones (Henderson-Sellars, 1979).

The initial treatment of water was often minimal or consisted only of filtration to remove bacteria and other particulates. It became more widespread, after the turn of this century, to deal with a variety of contamination problems from bacteria to animal infestations. For example, the Hamburg, Germany, water supply system had to introduce slow sand filtration in 1894 to eliminate animal penetration from the river Elbe and counter a cholera epidemic. Torquay, England, had to introduce filtration to an high-quality unfiltered supply which was found in 1910 to suffer from an infestation of the colonial, encrusting Polyzoa ('moss-animals') which blocked valves and water-meters (Smart, 1989). Disinfection of water supplies, usually using chlorine, began in Britain in 1905 and the United States in 1908 (Montgomery, 1985).

In older water undertakings filtration systems introduced over fifty years ago have been unable to cope with the effects of eutrophication of water supplies, the most common of which is blockage of filtration systems by algae (Scott, 1975). In the UK such problems became more frequent after the 1940s, with increasing frequency of problems and reduction of supply volumes. This is partly because the older upland supply catchments have become affected by more intensive agricultural and forestry activities, and partly because new reservoirs have had to impound more polluted river sources. Most of the reservoirs constructed during this time have had to be in the lowlands because of shortage of suitable new upland sites, often augmenting direct supply from the catchment with pumped supplies from neighbouring rivers. Treatment processes have increased in sophistication because the lowland river sources were inevitably contaminated with sewage effluent as well as enriched with nutrients. In the last two decades there has been increasing concern about contamination of water supplies from all sources, including groundwater, with dissolved organic contaminants (Montgomery, 1985).

5.2.2 Eutrophication effects on treatment processes

It is thus difficult to separate out the effects of eutrophication upon water treatment processes and costs from those of other contaminants. The most direct effects are those caused by increasing amounts of algal and zooplankton blockages and associated problems such as taste and odour in sources which formerly did not experience such problems. In the upland Talybont reservoir in South Wales, classified as 'meso-oligotrophic' (Scott, 1975), algal problems in the treatment process began in the early 1960s following small increases in phosphorus concentration which were caused by the fertilisation of new tree plantings in an afforestation scheme and the provision of sewage-treatment facilities for the increasing number of recreational visits to the catchment. The treatment process, filtration without coagulation, was unable to retain filaments of the dominant species, *Oscillatoria*, with the result that penetration into supply caused widespread complaints of taste and odour, poisoning of industrial ion exchange resins and a subsequent increase in animal infestation of the distribution system. In this and other smaller upland reservoirs, treatment costs directly increased through the need to dose the reservoirs with copper sulphate once or twice each year; they also indirectly increased through reduced output from shorter filter runs and subsequent distribution problems. In the longer term, further capital investment was necessary to up-rate the works capacity and introduce coagulation prior to the filtration stage.

There is limited evidence of similar long-term changes to the treatment of lowland water. Houghton (1972) reported that in treating water from the River Stour in Essex, south-east England, algal crops had progressively increased with the result that the filtration system employed (slow-sand filtration) produced only 60% of the water per unit area of filter bed than it did twenty years previously.

Treatment processes on lowland surface sources (rivers and reservoirs) have often been more elaborate for a long time. In the long term, technologies exist to deal with high quantities of particulate contamination. This has enabled, for example, the former Metropolitan Water Board (now Thames Water) to supply London with water from highly eutrophic sources since the beginning of this century with only four interruptions caused by algal problems. This overall success however, hides the problems which frequently arise in the short term as certain kinds of algal or zooplankton particle overcome certain stages in the treatment process.

The commonest problems are filter blockages, accounting for 80% of algal problems in Great Britain up to the 1940s (Anon., 1949). Most of the taxa causing blockage were, and continue to be, diatoms (Collingwood, 1977). Filter processes in Britain are either of two types, double filtration where a slow sand filter is preceded by a rapid sand or micro-strainer stage (DSF), or a coagulation/ sedimentation stage preceding rapid gravity filtration (CSF). In the DSF process the pore sizes in a slow sand filter bed are small, approximately 10 µm, because the effective filter is the sand together with a living community of microorganisms known as 'schmutzdecke' (Pearsall *et al.*, 1946). They are preceded by a rapid filtration process where larger particles penetrate sand grains with a pore size of 100–150 µm and are easily removed by backwashing, allowing the slow sand filters to run for long periods without cleaning. In the CSF process the larger pore size of the rapid sand filters works effectively because it is preceded by a stage where organic particles are coagulated into larger 'flocs' by addition of alum or ferric sulphate and a polyelectrolyte, and sedimented. There is also some 'carry-over' of floc particles onto the sand surface which reduces effective pore size and enhances particle filtration. A typical CSF works is illustrated in Fig. 5.1.

The differences in pore size and effective particle size between the two processes result in different operating responses to the same algae. Large algae, such as filamentous or colonial forms, are more effectively dealt with by DSF, which removes most particles in the first filter, permitting long filter runs in the second. Large algae are less easily coagulated than small ones however, so CSF has greater difficulty in dealing with them. At Farmoor reservoir, Oxford, England, removal of a crop of the filamentous diatom *Melosira granulata* (of 34 µg/l chlorophyll 'a' in the reservoir) by coagulation/ sedimentation was only 50% with the result that the filters became overloaded and runs dropped from 24 to 5 hours. Preliminary microstraining however, removed 78%, the coagulation/sedimentation a further 11%, allowing the filters to effectively remove the remainder (Youngman, 1975). Microstrainers conversely had negligible effect upon populations of the small centric diatoms *Stephanodiscus astraea* and *S. hantzschii* (7% removal) but these were effectively eliminated by coagulation/sedimentation (a further 89% removal) so that crops of up to 95 µg/l chlorophyll 'a' could be dealt with without reduction in works output.

Slow sand filters are more easily blocked by species such as those of *Stephanodiscus*, and problems have arisen with only 2,900

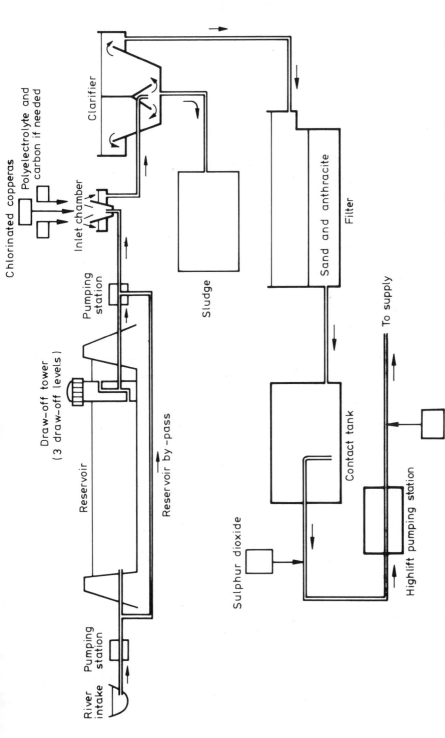

Figure 5.1 The layout of a typical CSF water treatment works, Melbourne, English Midlands. Modified from Brierley (1984), with permission.

cells/ml (compared with a maximum of 14,000 cells/ml at Farmoor) at Kempton Park treatment works, London (Taylor, 1963) which necessitated their closure until the reservoir population had declined.

Zooplankton have historically caused few water treatment problems because, being large particles, they are effectively removed by filtration and microstraining, if present. Recently however, problems were caused in Rutland Water, England, by populations of *Cyclops* spp. of up to 150 individuals/ml. These reduced filter runs by direct blockage because they were unaffected by the prior coagulation/sedimentation stage. Additionally the runs were reduced to such an extent that there was insufficient time for the filters to be backwashed or for the wastewater (the supernatant of which is normally returned to the head of the treatment works after 2 hours settlement (Ford *et al.*, 1982)) to be recovered because of the concentration of still-living zooplankton from the filters. Similar problems of zooplankton concentration in washwater recovery have occurred at Melbourne Treatment Works, Derbyshire; they can be overcome only by discharge of the washwater to waste, thus losing its original pumping costs which can amount to several hundred pounds sterling per day (1985 prices) (Barham, 1986). The only effective treatment has been found to be compressed-air flotation, whereby the zooplankton particles act as nucleii for air bubbles and float to the surface in a scum which can be skimmed off (Wilkinson *et al.*, 1981).

In general, DSF treatment is more effective in total removal of incoming algae, though prone to complete blockages, whereas CSF is usually effective in removal of a constant proportion of incoming algae. This means that, with high algal concentrations in the raw water, penetration of the whole CSF treatment into the distribution system is substantial even if it is only a few percent in relation to the raw water algal concentrations. Penetration of this organic matter leads to taste and odour problems associated with its decomposition by algae and fungi, but also provides food for the build-up of an animal community which itself causes quality problems. Animal infestation is rare or absent in groundwater-derived water supplies, uncommon in surface water supplies treated by DSF but widespread in supplies treated by CSF (Smart, 1989). The economic costs of this infestation lie largely in its control, which is usually practised by pyrethrin/permethrin dosing of the system or by periodic flushing at the access points (fire hydrants in the UK).

In a review of the effects of eutrophication upon the surface-water derived supplies of the Anglian water supply region of eastern England, which account for 35% of total supply, Hayes and Greene (1984) identified five impacts upon the supply processes, in addition to the consequences of physical penetration of treatment described above. These are:

1. Algal breakdown products, mainly mucopolysaccharides, are able to chelate the Fe/Al added as coagulants, leading to increased metal complexes passing through into supply. This is associated with green algae and cyanobacterial blooms. The complexes may then precipitate under low pH conditions causing industrial use problems such as in the production of carbonated water for the soft drinks industry.

2. Increased use of aluminium sulphate as the primary coagulant, particularly during times of high algal crops, results in higher levels of dissolved aluminium in supply. There are perceived links between aluminium levels and encephalopathy in renal dialysis patients, and EEC Directives now seek to limit the concentration of aluminium in tap water to 200 µg/l.

3. Oxygen depletion in reservoirs during warm summer months can occur following the collapse of algal blooms, even under isothermal conditions, leading to accumulation of ammonia, iron and manganese in the water column. Ammonia affects the oxidation and disinfection capacity of chlorine. Soluble Fe and Mn penetrate supply and cause discolouration problems (e.g. in washing machines) together with build-up of deposits within the supply network.

4. Haloforms, a group of volatile low-molecular weight halogen-ated organics, occur as a result of the reaction of chlorine with organic precursors principally derived from algal exudates. Levels of up to 200 µg/l have been recorded on occasion and levels are generally higher in the Anglian region than elsewhere in the UK. Public concern over these compounds, the commonest of which is chloroform, relate to their possible carcinogenic properties.

5. Taste and odour problems, either from algal penetration, from the breakdown of dense algal crops in raw water supplies, or from the growth of microorganisms in the littoral zones of reservoirs occur during elevated summer temperatures. They are usually dealt with by dosing treated water with high concentrations (up to 40 mg/l) of powdered activated carbon prior to a settlement stage. This is now widely used in the

treatment of lowland sources as a mechanism of eliminating trace organic compounds of industrial origin as well as alga-derived organics.

Hayes and Greene were unable to give any firm costs associated with the the problems of eutrophicated supplies, but an analysis of the increased costs of algal treatment was made for a eutrophic reservoir source in Japan by Magarara and Kunikane (1986). They showed that as algal cells in the raw water increased in concentration from 3,000 to 30,000/ml, conventional treatment needed to be supplemented by granulated active carbon filtration: between 30,000 and 300,000/ml this needed to be supplemented further by ozonation. The total extra costs of treatment for a plant producing 105 m^3/day potable water amount to a sterling equiv-alent of £19,000 at 1986 prices, excluding any additional manpower, pumping or higher chlorine dosing costs. They also excluded capital costs of plant installation.

An additional problem in water supplies is that of the direct concentrations of dissolved nitrates. Quite apart from their contributory effect with phosphates on plant growth discussed in Chapter 2, high concentrations of nitrates are believed to carry certain direct health risks to humans. The most understood, though fortunately very rare, effect is to cause a disease known as methaemoglobinaemia in infants under three months old. Nitrate is converted to nitrite which oxidises the ferrous iron of blood to ferric, forming methaemoglobin instead of haemoglobin, which is inefficient in binding oxygen. Five percent of haemoglobin conversion causes cyanosis, 30–40% causes hypoxia and over 50% is fatal. Only 10 cases of this disease have been recorded in the UK over the past 30 years, but they are usually caused by mixing up powdered milk formulations with nitrate rich water (Anon., 1983).

The other potential health risk associated with nitrates is their conversion in the digestive process to nitrosamine. These have been shown to be powerful carcinogens in animals, though with no direct evidence that human cancers are accelerated as a result of nitrate consumption. Moreover much human ingestion of nitrates is in foodstuffs rather than water (Anon., 1983). Nevertheless, the World Health Organisation established three ranges of nitrate levels in water supplies: unacceptable levels of over 100 mg/l nitrate (22.6 mg/l as N), acceptable levels of 50–100 mg/l nitrate and recommended levels of under 50 mg/l as nitrate. The EEC in 1980 (European Economic Community, 1980) established 50 mg/l nitrate as the maximum acceptable concentration for member states and half that as the guide level, beyond which bottled water supplies should be provided for infants. Highest concentrations, often above the acceptable levels, are usually found in groundwater supplies from areas of intensive arable agriculture because surface water storage in reservoirs provides effective nitrate reduction through denitrification (Chapter 3). Costs of nitrate removal through blending of sources, removal by ion-exchange, and loss of agricultural production through fertiliser use restrictions will be high in the UK; from £10 million to £80 million per year at 1982 prices has been estimated (Anon., 1983).

5.3 FISHERIES MANAGEMENT

Most lakes and rivers throughout the world are exploited for fisheries, either for direct human consumption or for sport. The effects of eutrophication in increasing the biomass and yield of fish stocks but changing the species composition were explored in Chapter 4. Increase in biomass may increase the value of fish, but changes in composition are almost always in the direction of lower economic value species. The history of the exploitation of fish stocks of Lake Erie illustrate this well, but it also illustrates the interaction of several other important factors besides eutrophication – principally other, more direct forms of pollution and of over-exploitation.

Lake Erie is subjected to intense human pressures: over ten million people inhabit its catchment area and sewage effluent from all the major conurbations on its shoreline enter the lake. The average concentration of nitrogen and phosphorus, and the average biomass of phytoplankton in lake waters increased three-fold between 1930 and 1965, but because of the greater duration of populations, phytoplankton productivity increased about twenty-fold (Regier and Hartmann, 1973). The effect of the subsequent increased rain of detritus on the lake bottom was to increase the sediment oxygen demand five-fold between 1953 and 1963 (Beeton, 1969), creating anoxic conditions over large areas of the lake bottom for several weeks during summer by the 1950s. These conditions spread to all three basins of the lake, developing in the eastern, deepest basin by the 1960s. Coldwater fish which normally inhabit the hypolimnion during summer, such as lake trout, lake herring, lake whitefish and blue pike thus lost their summer refuges.

In addition, the spawning areas of the western basin sediments have been lost because of the combination of low oxygen and deposition of fine silts. It is probable that eutrophication is responsible in these two ways for the decline in commercial catches of blue pike, lake whitefish and cisco, the former two particularly in the 1950s and 1960s (Beeton, 1969).

Many other pressures however, have also affected the lake fisheries. Since the latter half of last century, forest clearance and agricultural development have increased catchment soil erosion leading to siltation of shallow near-shore spawning grounds. This has been coupled with clearance of freshwater marshes, which once covered about 10,000 km^2, losing valuable juvenile fish sheltering and feeding areas. On the incoming rivers, mill dams impeded the spawning movements of species such as walleye and sturgeon.

Overfishing following the development of gill nets in the latter half of last century has also played a contributory role in the decline of individual species, although precise measurement of such effects has been difficult. The most prominent indication that this has taken place is the progressive change in catch composition from the most economically valuable species – lake trout and sturgeon at the beginning of this century – to less valuable species such as yellow perch and smelt at the present day, without significant change in total biomass harvested throughout the century.

Introduced species of fish have probably also contributed to the changing composition of catches. Smelt invaded the lake in 1931, becoming abundant by the early 1950s. These are pelagic fish which would normally be controlled by predation from species such as trout and blue pike. In the absence of such control, the high numbers of adult smelt themselves probably exerted significant mortality pressure upon the juvenile populations of other species.

Most recently, disposal of toxic wastes have resulted in elevated levels of such substances as mercury and persistent pesticides accumulating in fish tissue, mercury in particular exceeding statutory limits in some species (Laws, 1981).

These complex combined effects of man, compounded by the problems of co-ordinating several regulatory authorities (Canada and USA, with four US states), have caused difficulties and delays in the restoration of Lake Erie. Many of the human effects (e.g. physical modifications of shoreline) are in any case irreversible, making eutrophication control only one of the necessary restorative measures.

In the UK, few freshwater fisheries are exploited only for human consumption; game fishing for salmonids and angling for coarse fish are primarily recreational activities even if the former are consumed. In the estuarine and marine environment, fisheries for direct human consumption, including culture, are more usual. In all of these there is the same difficulty of separating out the effects of eutrophication from other adverse human influences, but certain effects can be examined.

The effects which algal blooms have on fisheries are almost certainly a result of nutrient-enhanced growth. There are many instances of the decay of blooms causing severe oxygen depletion in lakes resulting in fish kills, particularly in shallow lakes with rapid mixing (Ayles *et al.*, 1976; Barica, 1978). The toxic products of cyanobacteria after bloom collapse in lakes has also been associated with fish kills (Sirenko, 1980), such as mass mortality of roach in Llangorse Lake, South Wales, during the late 1960s (Edington and Edington, 1977). Of greater economic impact however, are the toxic effects of dinoflagellate blooms in estuarine and coastal waters (Paerl, 1988); in 1971 for example, a 'red tide' toxic bloom of *Ptychodiscus brevis* off the Gulf Coast states of the USA killed about 100 tons of fish per day leading to a loss of $20 million along the Florida coast alone. Fish kills have also occurred in coastal waters off California and several Far Eastern countries. Less dramatically, but also with economic consequences, blooms of certain cyanobacteria, notably *Oscillatoria agardii*, are associated with 'muddy odours' in fish such as bream (Persson, 1980).

In rivers with extensive growths of rooted angiosperms, whose increasing biomass over the past few decades has been associated with nutrient enrichment, such as the River Nene in eastern England (Brierley *et al.*, 1989), fish kills can occur as a result of the night-time deoxygenation by the plant biomass. Occasionally such mortality may also occur in less nutrient-rich rivers such as the Wye in Wales, where deoxygenation by the dense beds of *Ranunculus fluitans* caused large salmon mortalities in 1976 (Brooker *et al.*, 1977).

More subtle effects of eutrophication on fisheries are harder to detect. Most British lowland lakes are maintained as coarse fisheries in a semi-natural state and extreme effects of eutrophication are evident as a consequence of the whole ecosystem effects such as observed in the current state of the Norfolk Broads (Chapter 8, and Moss (1989)). Most reservoirs are managed as 'put-and-take' trout fisheries for brown trout (*Salmo trutta*) and rainbow trout (*S. gairdneri*) (Crisp and Mann, 1977). In an analysis of the angling records for Eye Brook reservoir, in the English midlands, Taylor (1978) found that angling success was not generally affected by algal density but brown trout fishing was poor when there were large amounts of diatoms, and deteriorated during heavy blooms of cyanobacteria and of the dinoflagellate *Ceratium*.

The costs of management of artificial trout fisheries are quite high. At Rutland Water, a new eutrophic pumped storage impoundment in the English Midlands (Harper and Bullock, 1982), establishment of the fishery in 1976 cost £600,000 (Langford, 1979). Annual running costs are primarily those of maintaining a hatchery to supply 80,000 yearling trout (Moore, 1982) together with boat maintenance and bailiffing costs. Anglers travelled an average of 100 km for each day's sport in 1980 (Moore, 1982) on top of the costs of the fishing permit (approximately £200 for a season ticket at this time) and boat hire costs. In 1989, during a particularly hot summer spell, the reservoir had to be closed to all recreational users after dense populations of cyanobacteria, notably *Microcystis aeruginosa*, produced toxins the ingestion of which killed several domestic dogs and grazing animals at the reservoir margins (Anon., 1990). The economic damage to the fishery was compounded by the loss of revenue from casual visitors to car parks, sailing activities and associated service activities.

Freshwater fishing activities in the United Kingdom, although primarily recreational, are undertaken by around 4 million people with a direct expenditure of £650 million and a further indirect expenditure of £530 million (Eltringham, 1984), so that disruptions caused by eutrophication may well have an increasingly important economic effect.

5.4 LAND DRAINAGE AND WEED CONTROL IN RIVERS

Eutrophication effects upon rivers increase the primary production and biomass of algae (under slow-flowing conditions (Müller and Kirchesch, 1982)) and probably rooted macrophytes, although there are no good before and after studies for enrichment effects upon macrophytes (Westlake, 1975) other than on changes in species composition (Haslam, 1978). Rivers in lowland agricultural areas are important in drainage and sometimes navigation, so weed control is extensively practised (Barrett, 1978). The costs of this are between £40 and £60 million per annum in the UK (Connor, 1985). Eutrophication is only one of several changing influences on weed-cutting policy, the others being land drainage and fisheries needs, so it is not possible to separate out the costs due to nutrient-stimulated enhanced growth. There is little doubt, but no hard evidence, that widespread eutrophication of lowland rivers in eastern and southern England has led to the development of high biomass of the algae *Cladophora glomerata* whose accumulated masses may suppress rooted macrophytes and cause flow restrictions in late summer. The clearance of this ubiquitous weed has almost certainly increased costs by an unknown amount.

Weed cutting is closely associated with dredging (Brierley *et al.*, 1989), which is silt removal to maintain channel capacity for flood prevention or depth for navigation. Increased siltation may be associated with effects of eutrophication but it is much more frequently associated with improved drainage of agricultural land and urbanisation (Brookes, 1988).

5.5 WILDLIFE CONSERVATION

The ecological effects of eutrophication upon wildlife conservation were examined in Chapter 4. Expressing these effects in economic terms is very difficult because wildlife has always been regarded as a 'shared' resource belonging to the human community in general. Economic values can only be accurately applied as the costs society or its agents are prepared to expend in scientific understanding, maintenance or restoration of that resource, together with the costs which individuals incur in the use or enjoyment of the resource.

In the United Kingdom the Government's agency responsible for wildlife conservation is the Nature Conservancy Council. Its activities are supported by a network of voluntary organisations such as the Worldwide Fund for Nature (WWF), Royal Society for Protection of Birds (RSPB) and Royal Society for Nature Conservation (RSNC), which together have a membership of approximately two million people (Moore, 1987). Eutrophication is but one aspect of wetlands conservation (Newbold *et al.*, 1986) which can only be viewed in isolation in lakes of conservation interest threatened by nutrient enrichment from rural land use changes. The direct ecological effects are loss of plant species. This has been documented in lakes such as Bosherton in Wales, Loch Eye in Ross and Cromarty, Scotland, Lochs Davan and Kinord in Deeside, Scotland (Newbold *et al.*, 1986), and Llangorse Lake in Wales (Edwards and Brooker, 1982). Loss of plants and concomitant loss of invertebrate species through altered habitat or oxygen regimes has occurred in Lochs Balgavies and the Lowes in Tayside, Scotland (Harper, 1986) and in the Norfolk Broads (Mason and Bryant, 1975), although in the latter additional factors are boat turbidity and habitat loss from land drainage activities (Moss, 1977; George, 1976). Only in the Broads can economic values be calculated as the costs of restoration measures, principally phosphorus stripping from major sewage works, which currently costs about £1 million per year, and sediment removal from individual broads where size and ownership considerations make this practical. Only two or three have been thus treated, at a cost of around £100,000 (1982 prices) for a small broad (Cockshoot Broad).

In certain limited circumstances there may be beneficial effects of eutrophication upon nature conservation. The nature reserve created at Rutland Water from 1974 onwards (Appleton, 1982) was primarily designed for birds and in this it has been spectacularly successful, achieving the status of an internationally important site for wildfowl in little over 10 years since creation. Part of its success is due to the input of tertiary-treated sewage effluent from the small town of Oakham into shallow bunded lagoons, which sustains high biomass of (limited species of) macrophytes and benthic invertebrates upon which a wide range of aquatic birds feed.

5.6 PUBLIC HEALTH HAZARDS AND NUISANCES

Some of the direct public health implications of eutrophication have been discussed above in relation to the problems of water supply and the development of toxic blooms of cyanobacteria or dinoflagellates. About twenty-five species of cyanobacteria produce toxins, of three kinds: neurotoxins, hepatotoxins and lipopoly-saccharides. The commonest are hepatotoxins, produced by species of *Microcystis, Oscillatoria and Anabaena*. Deaths of birds, mammals, amphibians and fish have been reported from around the world for over a century, but no direct human fatalities are known. Gastro-intestinal effects can occur on humans through ingestion of poisioned shellfish or fish, and cyanobacterial toxins can also cause gastro-intestinal upsets and skin rashes to swimmers in freshwater lakes through contact and ingestion of water containing scums (Skulberg *et al.*, 1984; Edington and Edington, 1986).

The summer of 1989 was particularly notable in Britain and

several other European countries for a number outbreaks of blooms of toxic algae in eutrophic lakes and reservoirs, which although not serious in terms of the magnitude of their effects, generated a great deal of publicity and public concern. The immediate cause was a mild 1988/89 winter followed by high 1989 summer temperatures and prolonged calm spells. These led to the development of floating scums of cyanobacteria, particularly of *Microcystis aeruginosa*. Accumulation of these scums at the water surface and along shorelines, and the subsequent concentration of toxins within the scums, caused the deaths of several sheep and dogs at Rutland Water, Leicestershire, and probably the illness of canoeists at Rudyard Lake, Staffordshire. The extent of the occurrence of toxic blooms during this summer, a review of the properties and effects of toxins, and an assessment of management strategies for their future control, has been published by the UK National Rivers Authority (Anon., 1990).

An additional indirect hazard is the possibility of outbreaks of botulism, which occurs in shallow water during warm weather and oxygen depletion caused by decaying organic matter, favouring the development of the bacterium *Closterium botulinum* whose neuro-toxin causes the disease in waterfowl after ingestion (Holah and McIver, 1982). Such conditions occur in the Norfolk Broads following deoxygenation of shallow waters in summer after algal bloom collapse, and they exist in the bunded lagoons of Rutland Water, though without any recorded outbreaks to date.

A nuisance, though not detrimental to health, aspect of eutrophic waters, is the development of adult insect swarms. These are particularly associated with the shallows of eutrophic stored waters and develop in particular climatic conditions: high temperatures which promote synchronous emergence and wind which concentrates the swarms (Ridley, 1975). Species developing in reservoirs are almost always non-biting midges of the family chironomidae or chaoboridae, although biting flies may develop along the strand lines of lakes in algal masses washed ashore. In rivers, swarms of mayflies, caddis flies (both non-biting) and blackflies (biting midges) have occurred in the UK and USA (Edington and Edington, 1986). Their development cannot exclusively be linked to enrichment alone since they are often associated with input of effluents which may also contain particulate food material (Hansford and Ladle, 1979). Control measures, notably the spraying of insecticides, have been used and weather forecasts provide some predictive tool for such preventative actions, but generally public information is the main method used.

5.7 OTHER RECREATIONAL ASPECTS

Recreational uses of eutrophic waters are widespread in the developed world simply because so many inland waters experience the consequences of enrichment. To a certain extent these consequences are tolerated, such as angling on water bodies prone to insect nuisances (Ridley, 1975); and sailing and informal shoreline recreation on water bodies experiencing planktonic or littoral algal growths, but severe if localised economic consequences result if the facilities have to be closed, as happened at Rutland Water and many other reservoirs in eastern England during the summer of 1989.

In the United States, less so in the UK, active lake management programmes, particularly the physical management of aquatic plants, are widely practised. These are associated with the recreational needs of the public – prevention of noxious decaying plant masses, deoxygenation and fishkills, facilitation of boat access – rather than as effective eutrophication control measures (Burton *et al.*, 1979). Costs were estimated as approximately $200/ha (1979 prices). Many lake and river sites (throughout the world) are affected by alien species whose growth is enhanced by nutrient enrichment such as the floating angiosperm water hyacinth (*Eichornia crassipes*) and the fern *Salvinia molesta* (e.g. Chapman *et al.*, 1974; Verhalen *et al.*, 1985; Joyce, 1985; Room, 1990) but whose nuisance values are largely a feature of rapid growth in a new habitat.

In all the examples discussed in this chapter, the effects of eutrophication are difficult to separate from other aspects of human impact upon the aquatic ecosystem. Costs of any one impact are difficult to separate from the others except at the point of specific control technologies, and lake or river restoration policies can succeed only where they address all impacts and their ecological effects in concert, rather than in isolation (Moss, 1989).

It is worth remembering that not all effects of eutrophication are negative, particularly in tropical regions. The main benefits derived from enriched waters are an increase in fish yields, the facility for using them as less technical waste treatment processes, the re-use of nutrients on land through irrigation, and the possible harvesting of aquatic macrophytes for fertiliser or fodder (Thornton, 1987).

— 6

Prediction and modelling of the causes and effects of eutrophication

6.1 INTRODUCTION

All scientific study attempts to understand the workings of the natural world through the formulation and testing of hypotheses by impartial observation and experiment. The initial generation of hypotheses may come from the analysis of a series of observations. Hypotheses are then used firstly to explain past and present events, and secondly to predict future events.

A series of related hypotheses which have been supported by experiment and observation may be brought together in the development of a body of knowledge about the workings of the natural world which provide theories – explaining a sequence of events – and classifications – 'ground rules' which explain how similar processes operate under different environmental circumstances.

Underpinning these processes of hypothesis-generation and testing, theory formation, the development of classification systems, and prediction, is the use of 'models'. I define models here as any conceptual device which can assist in our understanding of the patterns and workings of the natural world. At their simplest, models can be mental – such a series of events drawn in boxes with pencil and paper, using arrows to show linkage and/or cause and effect; or a series of subjective groupings of similar elements in a data set – such as lakes divided into classes by their dominant phytoplankton or macrophyte species, or by phosphorus concentration (Vollenweider, 1968). Subjective classifications represented the first historical attempts to understand the basis of lake fertility and biological productivity; examples were given in Chapter 1.

At the next level of complexity models utilise mathematical relationships to summarise more complex data sets. Statistical methods enable data to be examined for significant relationships. Simple correlation between two variables shows that they are connected (although not necessarily as cause and effect) and allows tentative prediction into the regions beyond the regression line fitted to the data, or of the second variable if only one is measured. More advanced methods, such as analysis of variance, are used to test the extent to which one factor, or a combination of factors, influences a specific set of results (Krebs, 1989).

Another form of data summary is the calculation of indices. An index is usually a single solution of a mathematical equation which compares the relative abundance or importance of different variables at a site, or between two sites. It is usually applied to measures of species' importance in different components of communities, as some form of 'diversity' or 'similarity' index (Southwood, 1966; Krebs, 1989) such as phytoplankton in lakes (Sullivan and Carpenter 1982) but it may also be based upon ecological measurements such as chlorophyll or total phosphorus concentration (Carlson, 1977). Indices can then be subjectively grouped or classified.

Wide availability of computers has allowed the use of models routinely handling larger quantities of data input usually takes the form of one or more two-way tables, or matrices, such as for sites and species. Analysis can then take place in one or more of three ways: direct gradient analysis which can display the distribution of organisms along environmental gradients; ordination which arranges species and samples in a two-dimensional space according to their similarities and dissimilarities; and classification which assigns data to classes or groups, usually (but not exclusively) in a heirarchical fashion (Gauch, 1982). Three dimensional matrices, such as sites × species × time, can be accommodated within a two-way matrix (e.g. Williams and Stephenson, 1973). Multivariate techniques enable patterns to be objectively produced which summarise large data sets. Prior to c. 1950 patterns (classifications) were subjectively produced by scientists using experience and intuition from smaller data sets, although this could be quite sophisticated.

All the forms of model above help to explain biological patterns of distribution in space and time. Further analysis of changes in time usually starts from a statistical correlation where a causal link between the two variables has been independently demonstrated and a model developed based upon the regression equation. This allows prediction and subsequent testing of the predictions, but it is usually limited to a linear pattern of change. This is a static model, such as the regression between spring total phosphorus in lakes and mean summer chlorophyll biomass, which enables prediction of future chlorophyll from known phosphorus (Sakamoto, 1966). A more widely used static model is the 'input–output' model originally developed by Vollenweider to predict lake phosphorus concentrations from catchment loadings and refined by successive steps.

When rates themselves change, some form of dynamic model based upon sets of differential equations, is necessary. A large number of different kinds of models at different levels of complexity have been developed and reviewed. All require powerful computing facilities and only some give better prediction than static models.

6.2 EXPERIMENTAL APPROACHES TO MEASURES OF EUTROPHICATION EFFECTS

Models require extensive observational and experimental data to construct and calibrate them, and one of the simplest and earliest developments of experimental work was the investigation of algal growth responses to nutrient enrichment in laboratory cultures. These have an immediate objective of investigating whether algal growth is nutrient limited in a particular time and place, but also provide the rates of growth under different nutrient conditions for calibrating models. Algal assay is usually achieved by growing a test species of algae in filtered lake water (Anon., 1971; Miller *et al.*, 1978) although natural populations of lake algae can also be used (Goldman, 1964). Successive samples are grown with small additions of nutrients and trace elements and growth recorded as cell numbers, volume or biomass as chlorophyll 'a'. The results provide insight into the nutrient or nutrients which are limiting to growth at the particular time of water sampling, and on the maximum growth potential of the species under test. A more sophisticated development of algal assay is the enrichment of populations *in situ*, in enclosures, using radiactive carbon uptake as a measure of photosynthesis and growth rates (Gerloff, 1969).

A more recent development of this algal culture approach has been to investigate the physiological indicators of deficiency. Reduction in parameters such as algal cell chlorophyll, protein and nucleic acid content has been used (Healey, 1973). Decreases in the proportion of cells in colonial algae such as *Asterionella formosa* under nutrient limitation (Tilman *et al.*, 1976) and in overall colony size of *Microcystis aeruginosa* under light limitation (Robarts and Zohary, 1984) have also been shown. Ratios of various cellular components measured in both algal cultures and in situ, such as N:P, phosphatase:ATP and ATP:carbon were used as indicators of phosphorus and nitrogen deficiency by Healey and Hendzel (1976,1980).

6.3 THE APPLICATION OF MODELS IN EUTROPHICATION ASSESSMENT AND PREDICTION

Chronologically, the development of models began with the study of the relationships between phosphorus inputs to lakes and in-lake phosphorus concentration (Vollenweider, 1968), and between in-lake phosphorus concentration and phytoplankton chlorophyll (Sakamoto, 1966). These led to the development and refinement of methods for estimating catchment inputs, which is the first step in understanding and predicting the behaviour of lakes experiencing enrichment (Reckhow and Simpson, 1980b) (Fig. 6.1).

6.3.1 Catchment inputs

The types of catchment land use contributing to diffuse and point sources of nutrients were reviewed in Chapter 2. Vollenweider estimated catchment export of nitrogen and phosphorus for a 'representative area' of Europe by combining the data available at the time for runoff with proportions of land use and human density (Vollenweider, 1968). This enabled him to conclude that more than 50% of phosphorus came from point-source inputs, but less than 50% of nitrogen. A large number of further studies were published over the next decade or so which measured nutrient losses from different land uses. Uttormark *et al.* (1974) and then Beaulac (1980) reviewed the literature available on nutrient export from different

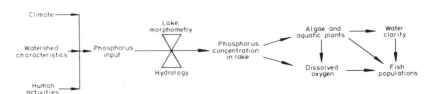

Figure 6.1 Simple conceptual model illustrating the factors controlling phosphorus loading and the measurable lake responses. Modified from Reckhow *et al.* (1980), with permission.

land uses, and Reckhow *et al.* (1980) presented a stepwise example of how export coefficients could be used to predict lake phosphorus loading, with subjective estimates of the accuracy of the prediction. With a large data base, the export coefficients for every different kind of existing (or projected) land use for any particular lake and its catchment could thus be derived from the most similar available data and the proportional area of the catchment. Reckhow *et al.* (1980) presented tables of export coefficients for nitrogen and phosphorus from raw data reported in the literature, 'screened' for accuracy (Table 6.1), and accompanied by guidelines for selection (Table 6.2).

The data sets were also summarised graphically as histograms and box plots which indicated the statistics of the sample set for each type of land use (Fig. 6.2). This presentation allowed the authors to recommend that users of the model procedure select 'most likely', 'high', and 'low' export coefficients for each land use type, summing them to arrive at three estimates of total annual catchment input to a lake for each nutrient. They provided a worked example of the procedure for Higgins Lake, Michigan, USA summarised in Table 6.3 (Reckhow *et al.*, 1980).

The ultimate aim of calculating and/or measuring catchment exports of nutrients is to obtain accurate data on the loading of nutrients to a lake, and the subsequent use which the lake ecosystem can make of these nutrients. Considerable care has to be taken in the design of any programme of investigation, and errors

Table 6.1 Criteria used in selection of export coefficients by Rechkow *et al.* (1980)

Accuracy	Good statistical sample design, careful use of controls
Precision	Sample strategy, number and frequency of samples
Representativeness	Comprehensive information on study catchments which would enable them to be easily matched with an unknown
Temporal extent of sampling	Only studies of one year and over were used
Concentration and flow data	Only studies which measured both were used

Table 6.2 Land use factors influencing the magnitude of export coefficients

Land use type	Factors to consider
Atmospheric input	Dustfall
	Soluble gases
Forest	Tree species type
	Soil type
	Bedrock and parent material
	Vegetation age
	Climate
	Disturbance
	deforestation
	forest fire
	fertilisation
Agriculture	
pasture and grazing	Rotational grazing
	Continuous grazing
	Fertilisation
	Animal density
crops	Soil type
	sandy/gravel
	clay
	organic
	Fertiliser application
	Tillage practices
	Crop types
feedlots	Percent impervious surfaces
	Animal concentration
	Covered feedlots
	Detention basin
Urban	Residential
	human density
	pet density
	vegetation percentage
	fertiliser use
	Public parks etc.
	Commercial/industrial
Sewage treatment plants	Type of plant
	Separate or combined
	Phosphate detergent bans
Rural septic tank systems	Fraction of year in use
	Detergent use in area
	Waste-flow reduction methods
	Phosphorus absorption of soils
	Soil drainage and permeability

Table 6.3 Example of calculation of catchment nutrient export from land use export coefficients

Step 1	Calculate inflow volume to lake (if not gauged)	$Q = (A_d \times r) + (A_o \times Pr)$
Step 2	Calculate areal water loading	$q_s = Q/A_o$
Step 3	Calculate areal phosphorus loading	$M = (Ec_f \times area_f) + (Ec_{ag} \times area_{ag}) + (Ec_u \times area_u) + (Ec_a \times A_o) + (Ec_{st} \times No.\ capita-years \times (1 - SR)) + PSI$

Where:

Q = areal water loading (m/ann)
A_d = catchment area (m^2)
r = total annual runoff (m/ann)
A_o = lake surface area
Pr = mean annual rainfall (m/ann)
q_s = areal water loading

M = total mass loading (kg/ann)
Ec = export coefficients (kg/ha/ann): f (forest), ag (agriculture), u (urban), a (atmosphere), st (septic tanks)
Area = area (ha) for above uses
SR = soil retention
PSI = point source input

can be minimised only by careful study of the problems encountered by other workers in their original publications. Bailey-Watts and Kirika (1987) for example discuss the problems of flow and concentration measurements in lake inflows from a variety of point and diffuse sources.

6.3.2 Lake loadings

The relationship between annual input of nutrients and in-lake phosphorus and nitrogen concentration is central to the understanding and management of eutrophication. Vollenweider (1968) showed that when catchment input to a lake was calculated as a loading per unit area of lake surface it was possible to compare lakes of different sizes. He tentatively concluded that the boundary between oligotrophic and eutrophic lakes lay at around 0.2–0.5 g total P/m^2/ann and 5–10 g total N/m^2/ann, based on data from 30 lakes whose trophic state was subjectively known from limnological observations. The actual concentrations in the lake however, depend upon a number of morphometric and hydrological factors as well as loading, the simplest and most important of which is depth. Vollenweider plotted phosphorus and nitrogen loading against mean depth and drew boundaries between oligotrophic and

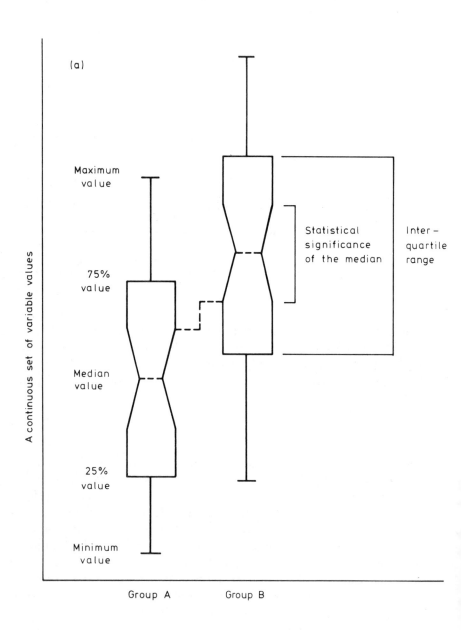

(b)

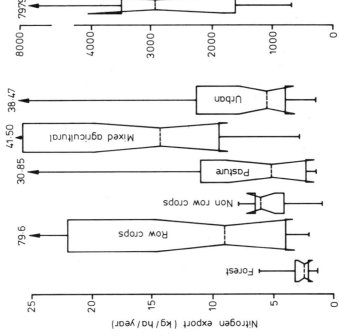

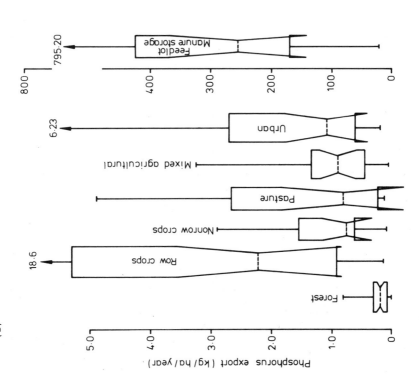

(c)

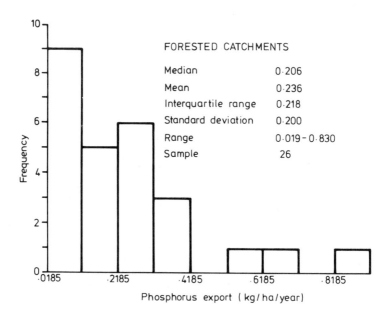

FORESTED CATCHMENTS

Median	0·206
Mean	0·236
Interquartile range	0·218
Standard deviation	0·200
Range	0·019 – 0·830
Sample	26

Phosphorus export (kg / ha / year)

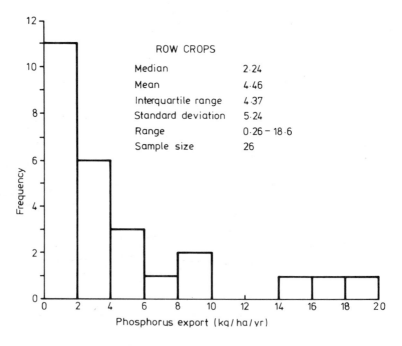

ROW CROPS

Median	2·24
Mean	4·46
Interquartile range	4·37
Standard deviation	5·24
Range	0·26 – 18·6
Sample size	26

Phosphorus export (kg / ha / yr)

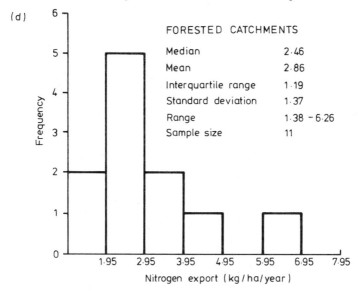

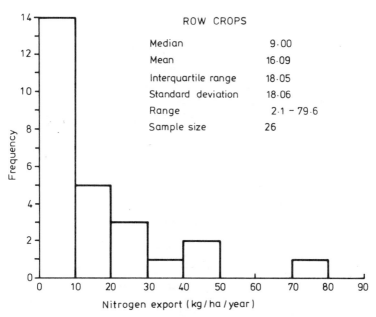

Figure 6.2 (a) Structure of box plots of nutrient export and their statistics. (b) Box plots of nutrient export from various land uses. (c) Comparative histograms and their statistics of phosphorus export from row crops and forested land. (d) Comparative histograms and their statistics of nitrogen export from row crops and forested land. Modified from Reckhow *et al.* (1980), with permission.

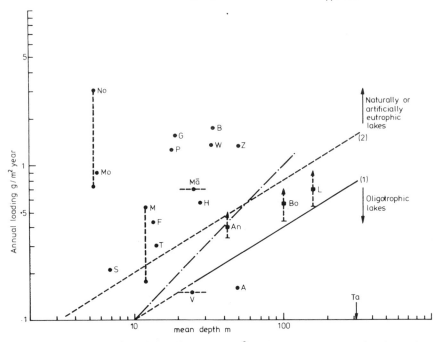

Figure 6.3 Phosphorus loading in g/m^2 ann against mean depth. Lakes are Aegerisee (A), Annecy (An), Baldeggersee (B), Bodensee (Bo), Fureso (F), Griefensee (G), Hallwilersee (H), Leman (L), Mendota (M), Malaren (Ma), Monona (Mo), Norrviken (No), Pfaffikersee (P), Sebasticook (S), Turlesee (T), Tahoe (Ta), Vanern (V), Washington (W), Zurich (Z). Modified from Vollenweider (1968), with permission.

eutrophic lakes; he then suggested 'permissible' loadings of N and P for different mean depths (Figs. 6.3 and 6.4). These loading graphs, improved by plotting loading against mean depth (z) divided by renewal time (T_w) (where T_w is volume/annual discharge) (Vollenweider, 1975), (Fig. 6.5), became the stimulus for much further work into the relationship between loading and in-lake nutrient concentrations, particularly phosphorus, and the biological consequences. Most of the subsequent modelling has focussed upon phosphorus, for several reasons. It is primarily derived from inflows (compared to nitrogen which has an atmospheric input *via* nitrogen fixation); it is almost always the limiting nutrient; its catchment sources are primarily point rather than diffuse; and therefore it is the nutrient most accessible to managed reductions. Dillon (1974b) reviewed the publications

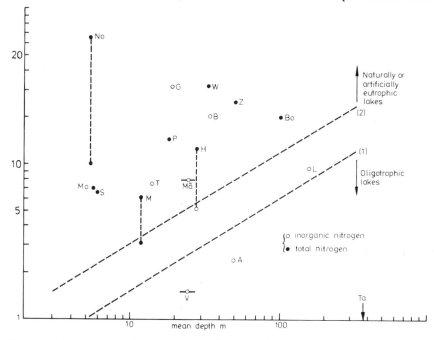

Figure 6.4 Nitrogen loading in g/m² ann against mean depth. Legend and origin as for Figure 6.3.

which at that time had used the two kinds of model to predict trophic status, and pointed out that the latter model gave a better degree of prediction. Lake Tahoe for example, an ultra-oligotrophic lake in the USA, with a very long retention time, fell much closer to the 'permissible' line when renewal was taken into account because a small increase in phosphorus loading would have a larger effect than the same loading on a more rapidly renewed lake.

6.3.3 Lake phosphorus concentrations

The fact that a graphical relation does exist between phosphorus loadings and a hydraulic function of a lake implies in the simplest case that there is a predictable relationship between loading and renewal time, and lake concentration.

Vollenweider showed for a small set of lakes (Vollenweider, 1968), that spring phosphorus concentration was indeed related to loading. A more recent refinement of this model, assembled from a data set of 87 lakes and reservoirs in the north temperate zone

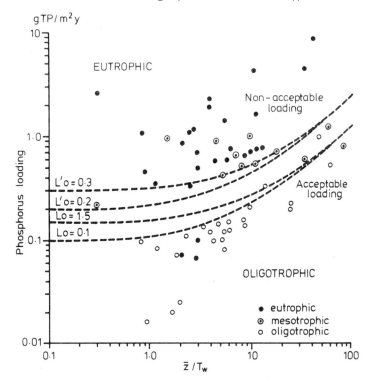

Figure 6.5 Phosphorus loading in g/m^2 ann against mean depth/renewal time. Modified from Vollenweider (1975), with permission.

(Anon., 1982), found the best predictor of average annual lake nutrient concentration to be the nutrient inflow concentration divided by a function of renewal time (Fig. 6.6). The assumptions of this relationship (Fig. 6.7) may lead to problems in some lakes, such as shallow lakes with an appreciable internal loading from the sediments, and these assumptions have been critically reviewed by Dillon (1974b).

Vollenweider developed a simple mathematical model for prediction of lake phosphorus incorporating these parameters, together with a factor for the sedimentation of phosphorus.

$$(P) = \frac{L}{z(\acute{O} + p)}$$

where (P) is total phosphorus concentration in µg/l, L is total phosphorus loading in gm/m^2/ann, z is mean depth in m, Ó is the

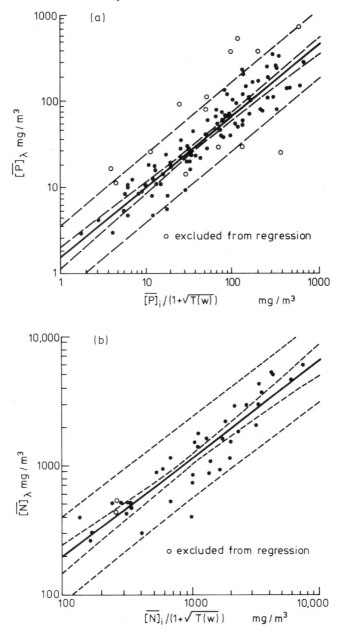

Figure 6.6 Relationship between average annual lake phosphorus (a) and nitrogen (b) concentrations and their loadings expressed as mean inflow concentrations divided by a function of renewal time. Modified from Anon. (1982), with permission.

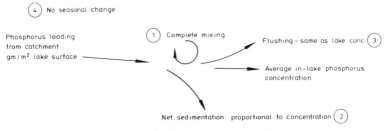

Figure 6.7 Assumptions of phosphorus loading – lake phosphorus concentration models.

sedimentation rate coefficient and p is the fraction of lake volume renewed annually.

The sedimentation coefficient is not easily measured, although it can be calculated from the other measured parameters by re-arranging the equation (Jones and Bachman 1976). They found a constant settling rate of 0.65 gave the best fit to data from 16 Iowa, USA reservoirs. Vollenweider however, developed the model to provide a more accessible measure of phosphorus sedimentation, the retention coefficient; that fraction of the total phosphorus entering the lake which is retained (Vollenweider, 1975). This can be theoretically related to the sedimentation rate coefficient and also is more easily measured, so that a modified equation can be used.

Vollenweider's modified model incorporating the retention coefficient, R, and the experimental measurement of R (R_{exp}).

$$R = 1 - \frac{p\,(P)}{L\,/\,z}$$

the measurement of R:

$$R_{exp} = 1 - \frac{q_o\,(P_o)}{\Sigma q_i\,(P_i)}$$

where q_o is the outflow volume in m^3/ann, (P_o) the outflow concentration, q_i the inflow volumes and (P_i) the inflow concentrations.

The equation then becomes:

$$(P) = \frac{L(1 - R_{exp})}{z\,p}$$

Dillon and Rigler (1974a) used the model to predict the spring phosphorus concentration for 18 lakes in Ontario, Canada and found for a subset of 13 lakes that the measured concentration closely matched the predicted concentration (r = 0.9) when shallow and high nutrient lakes were excluded. Kirchner and Dillon (1975) showed that phosphorus retention was related to areal water loading for these lakes, q_s m/ann, the outflow volume divided by lake surface area.

Chapra (1975) and Dillon and Kirchner (1975) showed for an extended data set that R could also be described by the 'apparent settling velocity', v, where:

$$R = \frac{v}{v + q_s}$$

Chapra derived a v of 16 m/ann, Dillon and Kirchner one of 13.2 and Vollenweider (1975) one of 10 (Fig. 6.8). Reckhow (1979b) subsequently proposed that v was related to q rather than just being a constant such that:

$$v = 11.6 + 1.2q_s$$

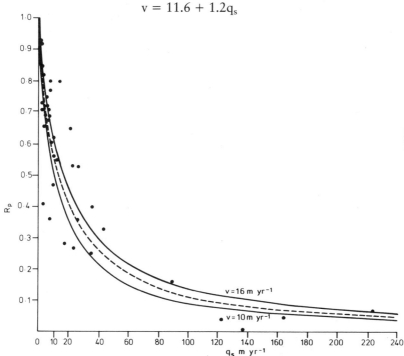

Figure 6.8 Phosphorus retention coefficient and areal water load with the regression line for $v/v+q_s$ where v = 13.2 m/ann. Modified from Kirchner and Dillon (1975), with permission.

Larsen and Mercer (1976) examined the relationship between R and a wider range of lake properties for a data set of 73 lakes in Europe and North America. They derived a best-fit value for v of 11.73, but suggested that R was better described by functions of q_s or p_w (hydraulic wash-out coefficient, outflow divided by lake volume), (Fig. 6.9). They also used a plot of (P_i) against R (Fig. 6.10) as an alternative to Vollenweider's phosphorus loading plot (Fig. 6.5) because the vertical axis, (P_i), deals with the inflowing processes (water supply and phosphorus loading) and the horizontal axis, R, deals with the processes of lake phosphorus assimilation. They argued that, although the relative position of lakes is the same on both kinds of graphs, using R provides more information about a lake's likely response to management changes.

Although the models predicting phosphorus concentration make certain simplifying assumptions, they have been found to be useful in a wide range of lakes. Osborne (1980), for example, studied a shallow lake in the Norfolk Broads with marked seasonal variation, high flushing rate and significant internal loading. He separated his analyses into three seasons – winter, spring and summer and obtained good predictions of mean lake total phosphorus using the

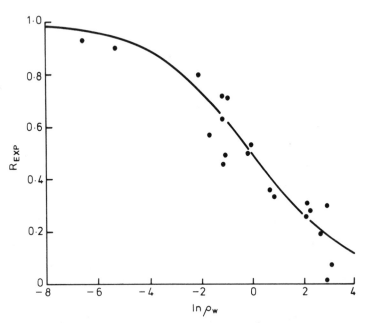

Figure 6.9 Phosphorus retention against a function of hydraulic washout. Modified from Larsen and Mercer (1976), with permission.

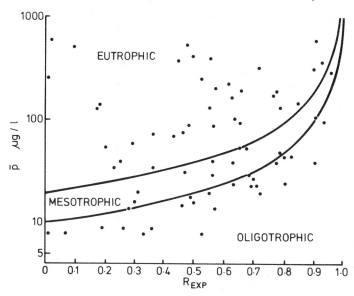

Figure 6.10 Mean inflowing phosphorus concentration against phosphorus retention coefficient for a range of lakes of different trophic status. Modified from Larsen and Mercer (1976), with permission.

Dillon and Rigler modification. Benndorf (1979) modified the model for reservoirs in the German Democratic Republic; he replaced total phosphorus by orthophosphate and introduced an outwash limit for very short residence times. He also found that it was necessary to use different values for v in eutrophic and oligotrophic reservoirs, 30 and 10 m/ann respectively, because of the vertical stratification and deep water outlet of the oligotrophic reservoirs.

A number of authors have made comparisons of different models with lake data sets. Mueller (1982) compared five model variants for reservoirs in the western USA and found the Dillon and Rigler (1974a) model to give the best fit. Hern *et al.* (1979), working with 39 lakes in the eastern part of the USA, compared the Kirchner and Dillon, Vollenweider, and Larsen and Mercier models; they found disagreement of 14%, 25% and 18% respectively between observed and predicted trophic state classification. Mamamah and Bhagat (1982) tested four models on 22 Western US lakes and found that even the best misclassified 23% of lakes. Working from a very large data set, of 704 natural and artificial lakes in North America and

Europe, Canfield and Bachmann (1981) tested 15 models and variants. They concluded that part of the prediction error of many models was due to inorganic particulate material in inflowing streams, sedimenting phosphorus at high hydraulic loading rates, whereas most models assume that phosphorus sedimentation decreases with hydraulic loading and increased lake flushing. The problem was greater in artificial lakes than natural ones. They used one subset of lakes to produce separate values for Vollenweider's sedimentation coefficient for natural and artificial lakes and the other subset to test them using his original model. They then compared the performance of the Vollenweider and several derived models with their own modifications which incorporated initial rapid sedimentation of phosphorus (by particulates near to the inflows) with subsequent rate either constant or varying with hydraulic loading. These latter modifications gave narrower confidence limits of estimation of lake phosphorus concentration, but even the best ranged between 31% and 288%. These examples illustrate that all the models need to be used with caution, but are an important first step in an analysis of lakes' eutrophication problems.

To help reduce the probability of wrong decisions arising from prediction error, Reckhow (1979a,b) and Reckhow *et al.* (1980) added a procedure for estimating the reliability, or uncertainty of these phosphorus models. These allow a user to estimate the error associated with phosphorus concentration predictions or the probability of a lake falling into a particular trophic state. This is based upon a first-order error analysis leading to calculation of confidence intervals. The problem of errors inherent in the use of these models is also discussed in Anon. (1982).

6.3.4 Phosphorus–phytoplankton and phosphorus–bacteria correlations

The graphical relationship between phosphorus and chlorophyll was one of the first to be observed and commented upon in the 1960s. Sakamoto (1966), in a study of the phytoplankton production of Japanese lakes of varying depth, showed a close relationship between mean chlorophyll content and both total nitrogen and total phosphorus (Fig. 6.11); Lund (1970) showed a similar relationship between winter maximum soluble phosphorus and maximum summer chlorophyll 'a' (Fig. 6.12). Dillon and Rigler (1974b) subsequently analysed a wider range of data on spring total phosphorus and average summer chlorophyll 'a' from Europe and

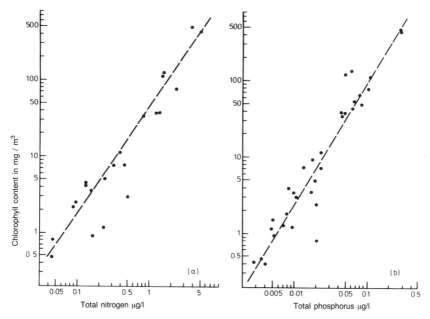

Figure 6.11 The relationships between lake phosphorus and nitrogen, and lake chlorophyll 'a' concentrations. Modified from Sakamoto (1966), with permission.

North America as a test of this relationship, and on finding close similarity further analysed the total nitrogen, phosphorus and chlorophyll from 19 Canadian lakes representing a variety of types. From their combined data, they calculated a regression which they suggested could be used to predict future mean chlorophyll from measured spring phosphorus:

$$\log_{10}(\text{chlorophyll 'a'}) = 1.449\log_{10}(P) - 1.136$$

Since then, a large number of studies and several collaboratively-collected data sets (e.g. the US Environmental Protection Agency National Eutrophication Survey, the Organisation for Economic Co-operation and Development) have produced similar regressions. The differences between many of the published regressions were examined by Nicholls and Dillon (1978) who attributed them to different analytical methods and sample timing. They also considered the factors influencing the varying chlorophyll concentration of algal cells, and from a study of the phosphorus–phytoplankton relationships in Ontario lakes, suggested that mean summer

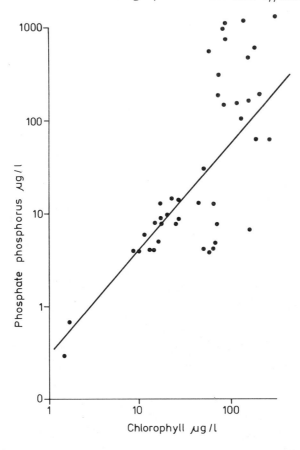

Figure 6.12 The relation between winter phosphate-phosphorus concentration and summer chlorophyll 'a'. Modified from Lund (1970), with permission.

phytoplankton cell volume was a more reliable biomass indicator than chlorophyll. Several authors have tried to refine the statistical relationship between phosphorus and chlorophyll by including nitrogen or nitrogen/phosphorus ratios (Smith, 1982; Canfield, 1983) in the regression. Sakamoto had previously pointed out that the yield of phytoplankton biomass depends upon the N:P ratio. Above 17:1 biomass was phosphorus-limited; below 9:1 it was nitrogen-limited. Other influences on the relationship include (for limited sets of lakes) area cover of aquatic macrophytes (Canfield *et al.*, 1984), which may compete with phytoplankton for nutrients.

The nature of the algal biomass has also been investigated, showing that with higher phosphorus, the proportion of nanno-plankton decreases (Watson and Kalff, 1981; Paloheimo and Zimmerman, 1983). Bacterial biomass has been related to chloro-phyll (Bird and Kalff, 1984) and here the regression suggests that bacterial biomass increases more slowly with phosphorus concentration than does chlorophyll. This could be due to higher bacterial production and loss processes, such as grazing, or that there is greater accumulation of phosphorus in algae in larger cells. The precise reasons require further investigation, but some have been suggested; Makarewicz and Likens (1979), for example, observed that the ratio of zooplankton to phytoplankton production increases in lakes with increasing trophy, and suggested that this could be because the zooplankton were making more extensive use of bacteria and detritus as food sources.

The relationship between measures of phytoplankton biomass and phosphorus becomes less valid as phosphorus concentrations increase. This was apparent in some of the earlier published graphs (e.g. Lund, 1970) but was not taken into account by the models (Dillon and Rigler, 1974b) which used lakes whose phosphorus concentrations were generally below 100 µg/l. Chlorophyll concentrations tend to plateau above a concentration of about 50–100 µg/l phosphorus, most likely because of self-shading. The lack of predictive power of linear models causes problems for phosphorus reduction strategies (Allan, 1980), but it has been overcome to some extent by the use of logistic models (Ahl and Wiederholm quoted in Ahlgren *et al.*, 1988; Straskraba, 1980) (Fig. 6.13).

6.3.5 Chlorophyll–transparency correlations

The biomass of algae is usually the major factor affecting transparency in lakes, at least during the growing season in temperate latitudes. Some studies have shown that artificial lakes have greater non-algal turbidities than natural lakes (Canfield and Bachmann 1981, for the USEPA lake eutrophication survey) and are better treated separately. Nevertheless, a strong relationship exists between transparency and biomass, which is usually expressed as Secchi disc extinction depth against chlorophyll (Fig. 6.14). The advantage of this is that Secchi disc extinction, being an easy and non-technical parameter to measure, has often been recorded for longer than chlorophyll and, with caution, can be used as an indicator of change in trophic state where other data do not exist.

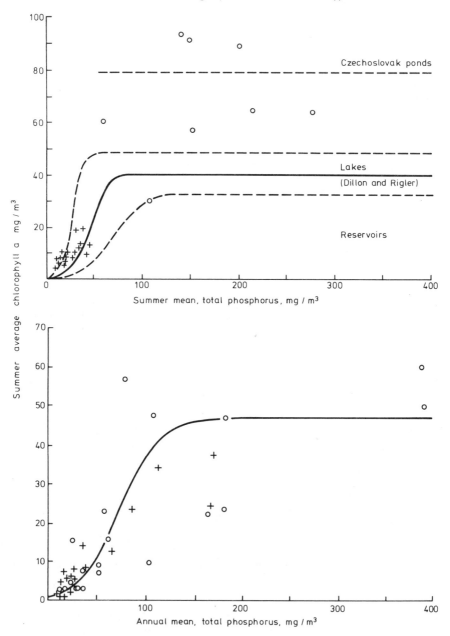

Figure 6.13 The plateau of chlorophyll 'a' with increasing phosphorus concentrations. Regression curves from North American data compared with Czechoslovakian reservoirs (+) and ponds (o). Modified from Straskraba (1980), with permission.

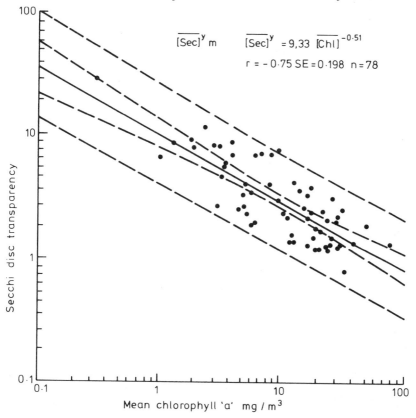

$$\overline{[Sec]}^y \quad m \qquad \overline{[Sec]}^y = 9,33 \ \overline{[Chl]}^{-0.51}$$

$$r = -0.75 \ SE = 0.198 \ n = 78$$

Figure 6.14 The relationship between chlorophyll 'a' and Secchi disc transparency. Modified from Anon. (1982), with permission.

6.3.6 Phosphorus–chlorophyll–productivity correlations

One would expect from the above discussion that phosphorus control of phytoplankton biomass over a large range of lake types would also extend to phosphorus control of primary production by phytoplankton. Some studies demonstrated such a link (e.g. Vollenweider *et al.*, 1974a), but on a large scale analysis of northern hemisphere results from the International Biological Programme, Brylinsky and Mann (1973) found that most of the variability in primary production was accounted for by latitude. In contrast, Schindler (1978) could not find a significant relationship between production and latitude but did obtain one with phosphorus concentration calculated from loadings, although he had

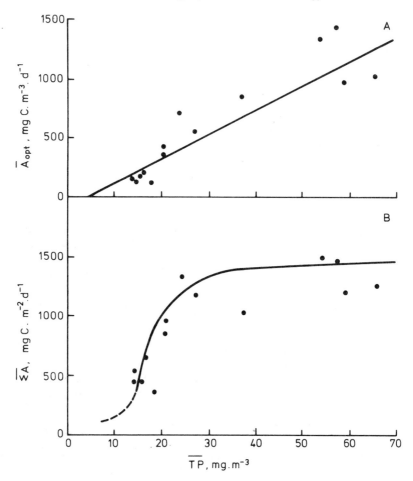

Figure 6.15 The relationship shown by volumetric photosynthesis (upper graph) and areal photosynthesis (lower graph) with total phosphorus concentration. Modified from Smith (1979), with permission.

few tropical or southern hemisphere lakes. In a more detailed analysis of 58 north temperate lakes, Smith (1979) showed that mean growing season photosynthesis per unit volume euphotic zone gave a strong linear relationship with both chlorophyll concentration and mean total phosphorus, and using a volumetric measure of photosynthesis avoided the plateau which appears in the phosphorus-area integrated photosynthesis relationship due to self-shading (Fig. 6.15).

6.3.7 Phosphorus–secondary producers correlations

All the main secondary producers of lake communities have been shown to have statistically significant relationships with measures of phosphorus and chlorophyll using the large data sets which have become available in the literature over the past two decades. McCauley and Kalff (1981) showed significant relationships between zooplankton (crustacean) biomass and phytoplankton biomass in a set of 17 lakes. The relationship suggested that zooplankton biomass increased more slowly than phytoplankton biomass across the trophic range, and they hypothesised that this was because nannoplankton biomass, which would represent the edible forms of phytoplankton, increased more slowly than total phytoplankton. Hanson and Peters (1984) also found the same slope to their regression, based upon 49 lakes from Europe, North America and Africa. Bays and Crisman (1983) showed in 35 Florida lakes that the proportion of microzooplankton to total zooplankton increased with trophic state, but Pace (1986) was unable to confirm this in a study of 12 lakes in Quebec.

Hanson and Peters (1984) also analysed profundal macrobenthos biomass against phosphorus, Secchi depth and several parameters of lake size for a smaller data set of 31 lakes. Their best regressions were against phosphorus, with a weaker (inverse) regression for Secchi, and weak or none for lake depth and area. Addition of lake surface area as a variable in a multiple regression made a significant improvement to the phosphorus–benthos relationship.

Fish yield has also shown a significant regression against chlorophyll (Oglesby, 1977) and total phosphorus (Hanson and Leggett, 1982). The latter authors also developed relationships for fish biomass against phosphorus, and against macrobenthos biomass/mean depth. In two separate data sets analysed, the best multivariate predictors included macrobenthos biomass, total dissolved solids, mean depth and lake area as independent variables in one, and total phosphorus, mean depth and total dissolved solids in the other.

6.3.8 Oxygen depletion models

An increase in biomass and production in lakes with increasing nutrient concentration, demonstrated above, inevitably produces more detritus when organisms die, and in stratified lakes, this detritus is mainly decomposed in the hypolimnion. Therefore one would expect to find greater oxygen depletion in nutrient rich lakes.

The volume of the hypolimnion is important however, providing the 'bank' from which oxygen is drawn during the period of stratification. Shallow lakes with the same productivity per unit area as deep lakes will develop oxygen-depleted hypolimnia more rapidly, but should show the same rate of oxygen depletion under a unit area of the hypolimnion. Initial attempts to prove this hypothesis foundered because the detritus loading was not measured. Lasenby (1975) was unable to correlate seston dry weight with hypolimnetic oxygen deficit, but did find a statistically significant negative correlation with mean Secchi disc transparency and explained this by assuming that transparency measurements incorporate the influences of both allochthanous and autochthanous particulate organic material, as well as dissolved organic matter. Cornett and Rigler (1979) used the Dillon and Rigler (1974a) model to calculate phosphorus retention as a measure of detritus sedimentation. They obtained a significant relationship with the areal hypolimnetic oxygen deficit, with mean volume-weighted temperature and mean thickness of the hypolimnion explaining most of the residual variation.

6.4 CAUTION IN THE USE OF REGRESSION EQUATIONS

The models presented above have a considerable amount of error associated with them, even though they are statistically significant. The plots are usually log-transformed to minimise the presentation of this error. There are adequate ways of estimating the confidence limits of any conclusions drawn from the regression, but such conclusions as are drawn have to be carefully used (Reckhow, 1979a). They should not be used to try to predict events for a single lake in any one year, because of the unpredictability of all the other factors − such as light, rainfall and temperature, which are seasonally variable. They can be profitably used in five main ways:

1. To predict an unknown variable from measurements of another in a low-intensity study or regional preliminary survey, e.g. mean chlorophyll from measured Secchi disc transparencies.
2. In cost−benefit analysis and similar management exercises, where the unit costs of, for example, phosphorus stripping of sewage effluent, can be weighed against the likely benefits in terms of chlorophyll biomass or transparency, and decisions taken about the level to which stripping can be afforded.

3. To guide further applied research. If a regression seems valuable, as were the original phosphorus loading and phosphorus–chlorophyll models, then this is a stimulus to further work which refines and tests it for different kinds of lakes in different parts of the world. The original phosphorus loading model of Vollenweider stimulated a great deal of research on the nature of phosphorus retention in different kinds of lakes, leading to improved versions of the original model. The phosphorus–chlorophyll regression has been confirmed (with minor differences) many times over, but refined for different lakes – such as those limited by nitrogen – as a result of research which shows a modification to the regression is necessary.

4. To generate further hypotheses about the nature of the regression. Its slope gives an important indicator about how rapidly one parameter is changing with respect to the other. The phosphorus–chlorophyll regression has led to much further work about the way in which the algal biomass changes with increasing phosphorus content for example, because it suggests that phytoplankton biomass increases at a greater rate than phosphorus. This is consistent with the observation that average algal size is larger at higher trophic state (Watson and Kalff, 1981) and cells have lower cellular phosphorus content (Shuter, 1978). In some cases further work may fail to support the original regression. Bays and Chrisman (1983), for example, found that there was an increase in the proportion of microzooplankton relative to total zooplankton with increase in trophic state. Pace (1986) confirmed the phosphorus–biomass relationship but found no evidence for change in community size structure. Other workers have independently found that species composition changes more or less predictably with increase in trophic state (Gliwicz, 1969) and further work will refine the predictability of the community structure.

When two models are compared they may lead to further speculation from which useful testable hypotheses develop. Comparison of the annual fish yield with fish biomass regressions suggests that fish in eutrophic lakes are more productive per unit biomass and therefore smaller. Comparison of the nannoplankton–total phosphorus regression (and assuming nannoplankton is primarily the edible algae for zooplankton) with the model for zooplankton specific grazing rate (Peters and Downing, 1984), allows grazing rate to be described as a function of phosphorus concentration. Since phosphorus concentration predicts zooplankton biomass and transparency, hence euphotic zone depth, the community grazing rate (percentage of the euphotic zone cleared per day) can be calculated and shown to rise with trophic state. Peters (1986) discusses these examples and deals with many other aspects of regression models.

5. To obtain a classification system, which on a large scale (e.g. country-wide) helps direct management decisions and focus resources for lake restoration where they are most needed. The principles of the major classification systems are discussed below, but in outline they all depend upon the definition of critical limits. Usually these define the trophic state, say for chlorophyll. For example, if a classification of mesotrophic is set for a lake, defined by its chlorophyll and determined by its uses, then the regressions enable one to calculate the level below which the lake phosphorus concentration needs to fall, and from that the level below which the phosphorus loading needs to fall, and from that identify the catchment sources of phosphorus most amenable to removal.

6.5 LAKE CLASSIFICATIONS BASED UPON CORRELATIONS AND LARGE DATA SETS

Vollenweider (1968) made an early approach to a quantitative classification of the trophic state of lakes by their phosphorus and nitrogen concentrations; based upon the chemical results of several earlier regional lake surveys he suggested a five-class tentative classification (Table 6.4).

Table 6.4 Trophic state classification by total P and N of Vollenweider (1968)

Trophic classification	Total P ($\mu g/l$)	Inorganic N ($\mu g/l$)
1. Ultra-oligotrophic	<5	<200
2. Oligo-mesotrophic	5–10	200–400
3. Meso-eutrophic	10–30	300–650
4. Eu-polytrophic	30–100	500–1500
5. Polytrophic	>100	>1500

Recognising that lake concentration is determined by nutrient loading, he converted these figures into 'permissible' and 'dangerous' (often now called 'critical') loadings, but qualified them by mean depth, recognising the work of earlier limnologists on the relationship between lake morphometry and productivity (Table 6.5).

Table 6.5 Vollenweider's loadings for total P and N $(gm/m^2/ann)$ for different mean depths

Mean depth up to	Permissible loading up to		Dangerous loading up to	
	P	N	P	N
5 m	0.07	1.0	0.13	2.0
10 m	0.1	1.5	0.2	3.0
50 m	0.25	4.0	0.5	8.0
100 m	0.4	6.0	0.8	12.0
150 m	0.5	7.5	1.0	15.0
200 m	0.6	9.0	1.2	18.0

These loading relationships have been refined by the development of the loading–mean depth/flushing rate models discussed above, but retaining the boundaries between 'permissible' (= oligotrophic below this level) and 'dangerous' (= eutrophic above this level) (Vollenweider, 1976) (Fig. 6.16).

The OECD (Anon., 1982) has extended the trophic state classification to include values for chlorophyll and transparency, based on the large data set accumulated by this co-operative study programme. Their proposed boundary values are given in Table 6.6.

Several other authors have proposed similar, but not identical categories (Fosberg and Ryding, 1980) and stressed the value of alternative variables for trophic state classification which can act as a check on the reliability of each other.

The OECD developed the classification further into an 'open boundary' system using the group mean values and standard deviation for each parameter, and the subjective classification of each lake in the data set by the limnologists who studied it. This system allows the uncertainty in allocating a lake to a particular

Table 6.6 OECD boundary values for trophic categories, in µg/l (except Secchi depths in m)

Trophic category	P	Chl.	Max. chl.	Secchi (m)	Min Secchi (m)
Ultra-oligotrophic	≤4	≤1	≤2.5	≥12	≥6
Oligotrophic	≤10	≤2.5	≤8	≥6	≤3
Mesotrophic	10–30	2.5–8	8–25	6–3	3–1.5
Eutrophic	35–100	8–25	25–75	3–1.5	1.5–0.7
Hypertrophic	≥100	≥25	≥75	≤1.5	≤0.7

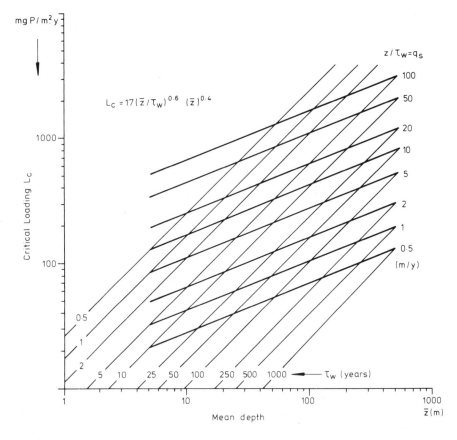

Figure 6.16 The lower limit of critical loading for phosphorus, based upon mean depth and renewal time. Modified from Vollenweider (1975), with permission.

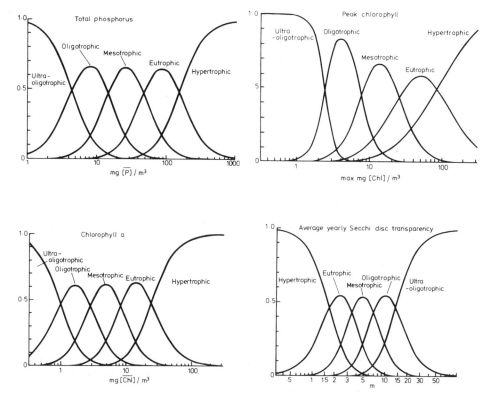

Figure 6.17 The OECD boundary system for lake classification. The curves express the probability of a trophic state classification based upon the parameter shown being correct. Modified from Anon. (1982), with permission.

category be taken into account when classifying it from measurements or calculated values of limnological parameters (Fig. 6.17).

The large OECD data base can also be used to predict unknown parameters using known ones, taking into account the confidence intervals calculated from the regressions equations. They recommend that only the 80% confidence intervals offer realistic predictive value, and that the diagrams should be used individually rather than sequentially to avoid magnifying the effects of the uncertainties.

An alternative approach to classification was developed by Carlson (1977), based upon the calculation of an index whose range was 0–100. The theory of the index is based upon the same statistical relationships between phosphorus loading, phosphorus concentration, chlorophyll and transparency discussed above, but uses the $\log_2$ of Secchi disc transparency as the starting point. Zero is set at 64 m, the integer greater than the maximum transparency ever recorded (42 m for Lake Masyuko in Japan (Hutchinson, 1957)), and each halving of transparency increases the index by 10. Chlorophyll and total phosphorus concentrations were related to transparency by regression equations and then added to the scale (Table 6.7). The trophic state index for a lake could then be calculated separately for all three parameters, or for one or two only. Figure 6.18 illustrates the index calculated as a yearly average for Lake Washington, which underwent progressive eutrophication during the 1950s and recovery during the 1960s. Carlson also tested it seasonally on lakes in Minnesota, and suggested that an index calculated from phosphorus values would be better in winter and spring, but from chlorophyll or transparency during the growing season.

An index of less than 20 represents ultra-oligotrophic, 30–40 oligotrophic, 45–50 mesotrophic, 53–60 eutrophic and above 70 hypertrophic (Kratzer and Brezonik, 1981).

This concept of a trophic state index has been developed in several ways. Huber (1982) modified it to use chlorophyll, rather than Secchi disc transparency as its basis for Florida lakes. Many of these are nitrogen-limited, rather than phosphorus-limited, because

Table 6.7 Carlson's Trophic State Index and its associated parameters

TSI	Secchi disc depth (m)	Total phosphorus (µg/l)	Chlorophyll (µg/l)
0	64	0.75	0.04
10	32	1.5	0.12
20	16	3	0.34
30	8	6	0.94
40	4	12	2.6
50	2	24	6.4
60	1	48	20
70	0.5	96	56
80	0.25	192	154
90	0.12	384	427
100	0.06	768	1183

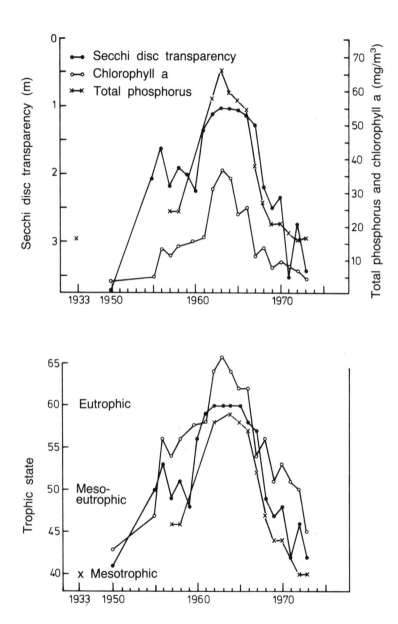

Figure 6.18 Carlson's Trophic State Index calculated sequentially for the restoration of Lake Washington. Modified from Carlson (1977), with permission.

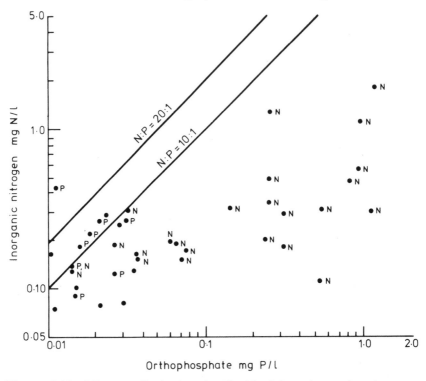

Figure 6.19 Nitrogen limitation in Florida lakes shown by their N:P ratios. Modified from Kratzner and Brezonik (1981), with permission.

of their underlying geology (Fig. 6.19). Kratzer and Brezonik (1981) therefore added a total nitrogen classification using a regression equation for chlorophyll and nitrogen and taking 10 μg/l chlorophyll as the boundary between mesotrophy and oligotrophy (TSI = 53). Ritter (1985) added a total nitrogen parameter and a dissolved oxygen parameter to the TSI for Delaware lakes and found that an average TSI for the five parameters offered a better classification than the original three-parameter index. The TSI of many lakes did not adequately describe their nuisance state however, because many suffer from excessive macrophyte and filamentous algal growths which alter Secchi disc transparencies. He suggested that some kind of total biomass parameter was needed, but did not propose one.

6.6 OTHER LAKE CLASSIFICATION INDICES

There are several other approaches to lake classification using the biotic community which have been developed as eutrophication indices. They are based upon two general hypotheses:

1. That species richness and diversity decline with increase in trophic state.
2. That dominant species or taxa in certain trophic conditions can be used as indicators for those conditions.

The discussion below will show that, in most cases tested, the first hypothesis cannot be supported, but the second has been, with some taxonomic groups.

The main groups which have been used for classification are algae (both phytoplanktonic and benthic), zooplankton, and benthic chironomids. Many authors have described changes in the biotic communities associated with eutrophication (discussed earlier in Chapter 4); only those taxonomic groups for which indices of some kind have been developed are discussed further here.

Phytoplanktonic algae have been investigated for longer than any other taxonomic group, and the major taxa were in use as a broad indicator by the 1930s; Pearsall (1924, 1930, 1932) showed that the change in algal composition was related to chemical factors in the water derived from the evolution of the drainage basin, and that there was a predictable progression from 'primitive' lakes dominated by desmids, through an intermediate type dominated by diatoms and desmids together, to the most 'evolved', silted lakes dominated by diatoms and cyanobacteria. Nygaard (1949) developed earlier European ideas of classifying these changes as four 'quotients', that is ratios of the number of species in different taxonomic groups with higher quotients indicating more eutrophic conditions. For example, the myxophycean (the old name for cyanobacteria) quotient was the ratio of myxophyceae/desmidaceae. A compound quotient summarised the effects of all four. Another approach which has been tested in eutrophication analysis is that of Palmer (1969) who developed lists of genera and species of algae, weighted by tolerance to organic pollution, from the results of a literature review.

The total number of algal taxa, usually identified to species, without weighting, have been used in ecological measures of species richness or diversity. The simplest measure of species richness is the total number of species recorded (Table 6.8). Many more complex

Table 6.8 Average number of species of phytoplankton in lakes of different trophic states in Ireland. From Round (1981)

Algal taxon	Oligotrophic	Mesotrophic	Eutrophic
Oligotrophic indicators			
Desmidacae	15.5	8.6	2.4
Chrysophyceae	3.3	2.8	1.2
Centric diatoms	2.7	1.8	0.9
Eutrophic indicators			
Cyanobacteria	3.1	5.2	7.2
Pennate diatoms	3.7	5.4	5.6
Constant taxa			
Chlorococcales	2.5	1.8	3.5
Dinophyceae	1.4	1.4	1.3

indices of diversity exist which relate the number of species to the abundance of each (Krebs, 1989).

All of the above methods suffer from a number of disadvantages. Brook (1959, 1965) has pointed out the difficulties associated with taxonomic separation of species of desmids and determining what is planktonic and what is swept in from the littoral. Any method which requires identification of specimens in a sample to species level depends upon the taxonomic skills of the investigator, which have to be particularly good with a water sample which could easily contain in excess of 100 phytoplankton species. Sullivan and Carpenter (1982) compared fourteen trophic state indices, including all the above-mentioned types and found only two to be of any use. Diversity indices showed too much variation and required too much time and effort; quotients were often unusable as the denominator was zero if no members of that taxon were found. The two of some use were Palmer's genus index, which showed the abundance of eutrophic indicator genera, and a concentration measure of the most abundant taxon, which was sensitive to the development of bloom-forming conditions.

In a more extensive comparison, Lambou *et al.* (1983) compared twenty nine trophic state indices, including Carlson's regressions, phosphorus loading models, algal assay techniques and the range of phytoplankton indices, in forty-four lakes. They tested their ability to rank the lakes in trophic state against rankings by measured total phosphorus and chlorophyll. They concluded that methods based upon phytoplankton distribution of community structure were particularly ineffective, and that the better indices overall were the phosphorus loading and Carlson-type TSI.

Despite the lack of widespread success for algal classification systems, they may have more restricted, regional use. Coesel (1975) has described eleven stages of trophic change based upon the desmid assemblages of waters (not just planktonic forms) and their dominant species. This kind of approach has also been developed for the benthic chironomid midge larvae of lakes. Benthic choronomids, like algae, have been used both for lake trophic state classification and for organic pollution assessment, usually in rivers. Indices have been developed using weightings for different indicator species, such as the 'Benthic Quality Index' of Weiderholm (1976), which correlates well with the ratio of average total phosphorus/mean depth of Swedish lakes. Sæther (1979) has pointed out though, that if the number of members of indicator communites was good enough to cover all forms of lake and be subdivided into enough compartments, then there is little need for an index. He has proposed fifteen characteristic chironomid communities, lettered consecutively in the Greek alphabet, six corresponding to oligotrophic states, three mesotrophic and six eutrophic, and developed a key to separate them (Sæther, 1979). The communities are strongly correlated with both total phosphorus/mean depth and chlorophyll 'a'/mean depth for a range of North American and European lakes (Fig. 6.20).

6.7 DYNAMIC MODELS OF LAKE ECOSYSTEMS

One of the limitations of empirical models is the difficulty in predicting the likely time of a lake's response to change, although they predict the probable end result, e.g. of phosphorus concentration with reduced loading. Some idea of the lake's likely response time to reduced loading may be obtained from the phosphorus residence time, calculating the new steady state phosphorus concentration from renewal time (Anon., 1982) (Fig. 6.21).

Dynamic models of lakes, or parts of lakes, differ from empirical models in that they set out to simulate rates of change between defined compartments within a lake. At the simplest level, models with two compartments were developed to deal with phosphorus exchange between lake and sediments. Chapra (1977) developed a simple dynamic model derived from Vollenweider's phosphorus

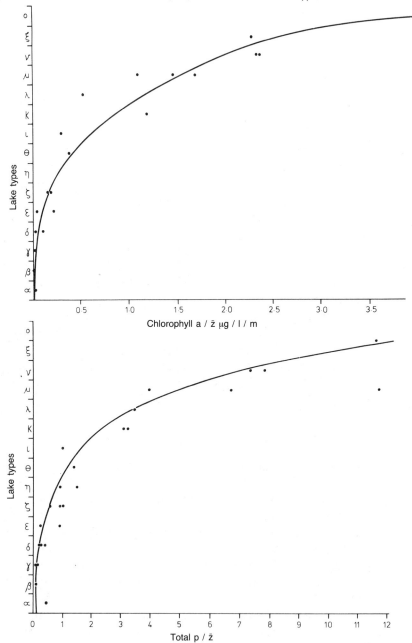

Figure 6.20 The relationship between chironomid indicator communities with phosphorus (upper graph) and chlorophyll concentration (lower graph) in a range of lake types. Modified from Sæther (1979), with permission.

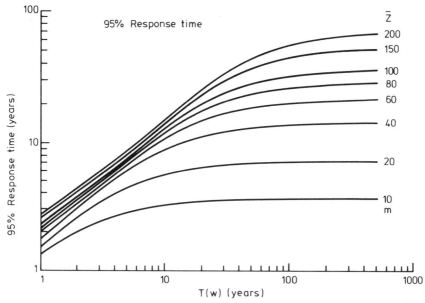

Figure 6.21 The calculated response time to changed phosphorus loading based upon mean depth and renewal time. Modified from Anon. (1982), with permission.

loading model which calculated lake phosphorus concentrations from loading, volume depth and hydraulic renewal as a function of time. Models such as this have given more accurate estimates of rates of change in phosphorus concentration after load reductions than Vollenweider's (Ahlgren, 1980). In stratified lakes, it is necessary to consider three compartments – epilimnion, hypo-limnion and sediments – with phosphorus in particulate and dissolved form, and exchange between them. Imboden (1974) developed such a model solved by four coupled linear differential equations to predict the steady state summer conditions and the hypolimnetic oxygen deficit, based upon phosphorus loading, exchange between compartments, photosynthetic and mineral-isation rates. Its subsequent development (Imboden and Gachter, 1978) (Fig. 6.22), using improved photosynthesis and hypolimnetic mixing terms, was used to predict the optimal eutrophication management strategies for Lake Baldegg, Switzerland (Imboden, 1975). It was considered successful, but encountered difficulties with large internal loading of phosphorus from the sediments.

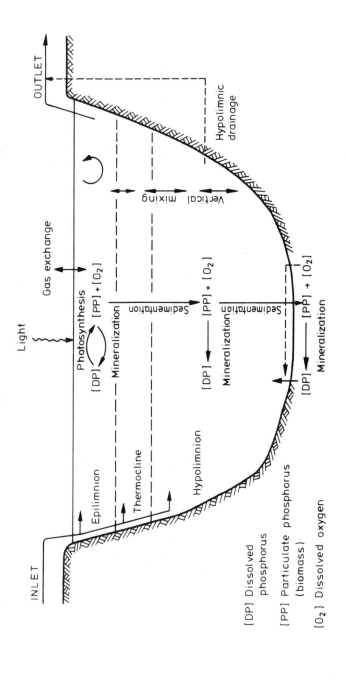

Figure 6.22 Concept of the dynamic model used to predict the effects of restoration on Lake Baldegg, Switzerland. Modified from Imboden and Gächter (1978), with permission.

Imboden's model assumed only phosphorus and light limitation of photosynthesis. In practice photosynthesis is limited by different environmental factors during the course of the growing season. Riley (1946) successfully modelled the seasonal development of phytoplankton by developing a differential equation describing population growth as a function of temperature, solar radiation, extinction coefficient of the water, depth of the euphotic zone, nutrient concentration and zooplankton biomass. She followed this with a seasonal zooplankton model based upon assimilation of available phytoplankton, respiration, predation and other mortality factors (Riley, 1947). In the decades since these simple models, accumulation of experimental data on environmental control of photosynthesis, and on phytoplankton–zooplankton interactions, has been considerable, permitting more sophisticated models to be developed.

Straskraba and Gnauck (1985) described the development of a series of models, AQUAMOD 1–3, which comprises only three compartments (or state variables) – phosphate, phytoplankton and zooplankton – but with a progressive increase in the number of feedbacks, such as self-shading on photosynthesis and return of phosphorus via mineralisation and sediment release. Benndorf and Recknagel (1982) also used a model, SALMO, with few state variables and many feedback control mechanisms to predict the effects of management options in East German reservoirs.

Numerous more advanced models have been developed, usually for specific lakes but some with more general applicability. A general model, called CLEAN, was developed in the USA, initially based upon 13 ecological state variables (Fig. 6.23) whose fluxes were described by 28 differential equations (Park and O'Neill, 1974) and linked to submodels for water balance and circulation patterns. The final version, MSCLEANER, had 31 state variables. It has been applied to a wide variety of lakes with some success (Fig. 6.24). A more detailed model of similar complexity (17 state variables) but with more detailed description of phosphorus dynamics (Fig. 6.25), was developed for Lake Glumsø in Denmark (Jorgensen, 1976) and validated by its predictions of the effects of sewage diversion in 1981 (Kamp-Nielsen, 1985).

Lake modelling from another approach, the thermal and hydraulic behaviour of lakes, was developed initially independent of water quality considerations. This treated lakes as a series of interacting vertical layers and using the principles of conservation of heat and of mass, allowed the successful simulation of stratification

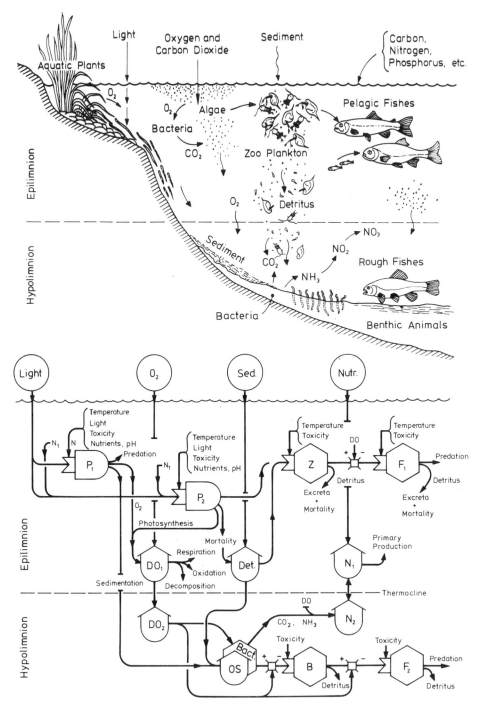

Figure 6.23 Concept of the ecological relationships used to derive the model CLEAN. Modified from Orlobb (1988), with permission.

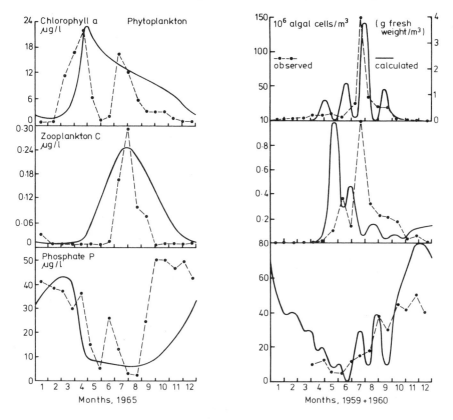

Figure 6.24 Comparisons of the predictions of the models AQUAMOD and MSCLEANER at Slapy Reservoir, Czechoslovakia. Modified from Straskraba and Gnauck (1985), with permission.

patterns (Orlob, 1988). Experience of this led to the inclusion of elements for nutrients and biota and the development of a model LAKECO which simulated the effects of altered hydraulic and nutrient inputs to Lake Washington (Chen and Orlob, 1975) upon algal biomass and distribution.

More advanced models consider multi-dimensional aspects of the hydrodynamics of lakes, horizontal as well as vertical patterns. Di Toro *et al.* (1975) designed a model of nutrients, phytoplankton and zooplankton for seven segments of Lake Erie; they calibrated it with historical data and speculated with it on the effects of different future changes in lake management. Chen and Smith (1979) developed a three-dimensional linked hydrodynamic–

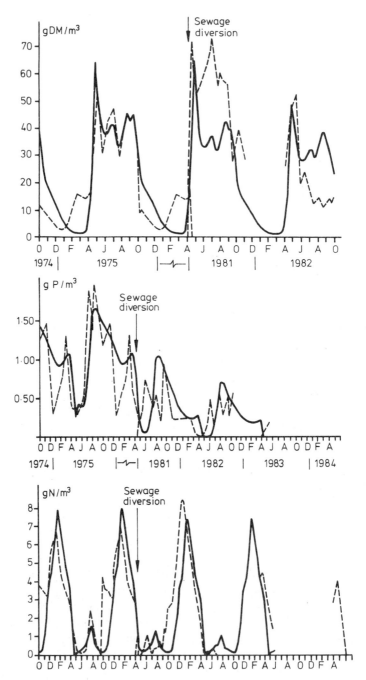

Figure 6.25 Predictions of the Glumsø model before and after sewage diversion (predictions = solid line, measured = dotted line). Modified from Kamp-Nielsen (1985), with permission.

ecological model for Lake Ontario over a one-year period with reasonable success. Theoretical models beyond this complexity are technically capable of development, but with so many state variables and interactions, run beyond the power of experimental results to calibrate them.

Many models have not been developed just for prediction and management, but often to try to summarise existing limnological knowledge in order to identify important areas of future research needs, particularly about rate processes such as sediment phosphorus return. Scavia (1979) gave an overview of such areas based upon the mismatches between modelled and observed events, particularly for the model MSCLEANER.

The use of models has brought out the two extremes of opinions about their effectiveness in advancing our understanding of limnology in general and eutrophication in particular. Statements that modelling is the only way to understand eutrophication (such as two quoted in Ahlgren *et al.*, 1988) are as unrealistic as the converse (Fryer, 1987). Contrasting views are explored in more detail by Peters (1986) and Lehman (1986). The important issue is how models can be sensibly used with carefully collected observational and experimental data – by understanding the limitations which were inherent in their construction and an awareness of the errors of their operation. Then they have a valuable role in guiding sensible management decisions and cost-effective research.

6.8 PREDICTION WITHOUT MODELS

It is possible to make predictions about the effects of future changes in the supply of nutrients to lakes without resorting to models, from experiments where the supply of nutrients can be altered as required. This approach is limited to simple laboratory or lake enclosure systems, and has only really been developed to predict algal growth potential as cell concentration, biomass or primary production because of the complexity of adding other trophic levels. Such algal bioassays have the advantage of replicability, and standard procedures for algal assay have been developed by the US Environmental Protection Agency using the species *Selenastrum capricornum* (Round, 1981). The techniques are versatile, so that they can be used for predicting the effects on fertility of water transfer and mixing schemes (Collie and Lund, 1981) as well as identifying limiting nutrients and effects of nutrient input reductions to lakes. They can also provide some of the quantitative data needed to calibrate models of growth and loss rates of phytoplankton (Uhlmann, 1971). They have to be used with caution because of their simplicity and usually are developed in conjunction with field programmes.

-7

The reduction of causes and the management of effects of eutrophication

7.1 INTRODUCTION

Extensive attempts to manage the adverse effects of eutrophication have been implemented in developed countries of the world for the past three decades. These vary between countries and between lakes depending upon the strength of public opinion, political will to enact legislative countermeasures, and of course the scale of the perceived eutrophication problem in the first place. Thus in Great Britain, eutrophication was not initially perceived as a serious problem (Lund, 1972) because most water supply treatment works were able to cope with the algal problems of eutrophic lowland waters, and no formal legislation exists to control nutrients. In the unusually hot summer of 1989 however, the widespread occurrence of toxic floating scums of cyanobacteria galvanised the general public, press, Parliament and the newly re-organised water industry into a frenzy of concern resulting in the closure of many water bodies to public recreation such as fishing and boating. In contrast, European countries such as Sweden and Switzerland have legislated to restrict both detergent phosphorus content and effluent phosphorus concentration; similar legislation exists in America and South Africa. Current developments of legislation on nitrate concentrations in water supplies in the European Community are the result of concern over the direct health implications of nitrates rather than their effects on eutrophication.

Eutrophication control strategies can be divided into two main areas – those concerned with the reduction of nutrient loads to lakes or in-lake concentrations, particularly of phosphorus, and those concerned with managing the existing high-nutrient state within lakes to minimise the adverse biological and chemical effects.

7.2 REDUCTION OF NUTRIENT INPUTS TO LAKES

7.2.1 Land use (diffuse nutrient sources) management

Most diffuse sources of elevated nutrient flows to watercourses are agricultural, although urban runoff through storm drainage is a significant and often diffuse source. A variety of agricultural practices contribute to nutrient runoff, and some of them are more amenable to control than others. It is convenient to distinguish three kinds of practice:

1. The nature of a crop, its sowing and harvesting
2. The application of fertilisers
3. The management of animals and their wastes

These practices differ in their relative importance for nitrogen and phosphorus runoff because of the different chemical behaviour of compounds of the two elements already referred to in Chapter 2. Nitrogen compounds are predominantly soluble and leached by rainfall whilst phosphorus compounds are predominantly particulate, bound to eroded soil particles. Phosphorus compounds are proportionately more concentrated in animal wastes, and more of the total phosphorus is in a soluble form.

The nature of agricultural crops has, until only recently, been determined by the economic decisions of farmers with little or no consideration of the effects upon the waters receiving runoff. In the past few years, the production of grain 'surpluses' in the northern hemisphere has led to economic and political controls on the crops which farmers may grow, and since these controls lead to lower intensity agricultural use (known as 'set aside' in the UK), they may result in lower nutrient runoff. Most 'set-aside' in the UK will result in the conversion of previously arable land to some form of grassland, with a reduction in nutrient losses both from the soil and from the cessation of fertiliser application. The reduction may only be short-term, because eventual ploughing of the grassland will release elevated quantities over a short period. No decisions have yet been made to control agriculture on a river or lake catchment basis although the opportunity may now exist to do this. Long term catchment management practices for control of diffuse sources nutrients in the drainage basin are possible. Peterjohn and Correll (1984) for example, showed that riparian forests in an agricultural catchments retained 89% of nitrogen and 80% of phosphorus received from arable cropland. Using 'set-aside' land in river valleys could give substantial nutrient removal rates in otherwise intensive arable areas.

The most important determinant of nutrient losses from different crops is the frequency and duration of the exposure of bare soil to erosion and leaching. Agricultural practices developed to control topsoil erosion, particularly in susceptible areas such as the mid-western United States and eastern England can reduce soil erosion by over 50% (Amemiya, 1970), and with it reduce particulate phosphorus erosion and surface nitrate leaching. Such practices include contour ploughing – crop rows oriented on the contours of the land with only a slight gradient towards the watercourse; mulch tillage – application of a surface layer of mulch (composted vegetation) and the ploughing-in of stubble from the previous crop; minimum tillage – to avoid compaction of the soil; the use of catch crops – fast-growing crops such as clover or alfalfa in between the harvesting and replanting of the main economic crop (Martin *et al.*, 1970; Addiscott, 1988). The use of trees and hedgerows as wind breaks will cut down wind erosion of susceptible soils.

Some of these techniques are not always compatible with modern farming methods. Hedgerows have been progressively removed in arable areas in order to allow large modern machinery to work more efficiently. Terracing on steeper slopes is not possible with large machines. There is still a conflict between the economic demands placed upon agriculture and their downstream water quality consequences.

The use of fertilisers on crops exposes a similar dilemma to other land use practices. Fertilisers have to be applied in one or more large doses, whereas crop demand is continuous during the growing season. Fertiliser application is now at the top of the parabola of diminishing returns (Bolin and Arrhenius, 1977) and is often applied in excess of plant needs through imprecise use and planning. Absolute reductions in the quantity of fertiliser used are possible, and the UK Water Act, 1989, makes this possible by the statutory declaration of 'nitrate-sensitive areas' where fertiliser use can be restricted legally. Runoff losses can be further reduced by correct timing of applications based upon crop needs rather than other farm management considerations such as labour and equipment availability, particularly by application at the beginning of the growing season rather than the previous autumn/winter when crops such as wheat are now frequently sown.

There are a number of future developments which could significantly reduce nutrient losses from arable agriculture (Anon.,

1983). The use of slow-release fertilisers in pelleted form is one, coupled with more accurate predictive models which can be used to produce optimum crop response to fertiliser application. New plant varieties which make more efficient use of soil nitrogen, and the opportunities offered by genetic engineering to exploit biological nitrogen fixation by its incorporation into crop plants or new symbioses is another. Enhancement of the microbial processes which convert organic nitrogen into available inorganic forms could make practices such as ploughing-in of stubble, and the application of animal wastes and sewage sludges more efficient and widespread.

Animal wastes and silage production for the winter feeding of animals contribute both diffuse and point-sources of nutrients to watercourses. The diffuse contribution arises through the spreading of slurry as an organic fertiliser addition, usually to grassland rather than arable crops. The timing of such application is important, because substantial losses of dissolved nitrogen and phosphorus occur if this is made in the winter, particularly on frozen ground (Porter, 1975). The larger problem is that of point source animal wastes, usually because animal production is centred upon many indoor units, with feed brought in as crop plants from outside, and limited opportunities for application of the wastes to land as organic fertiliser. Currently there are disincentives to the storage and re-use of animal wastes: their high volume relative to inorganic fertilisers because of their water content, their low proportion of nitrogen relative to phosphorus, and their noxious smells which encourage disposal rather than storage for future use. The relatively low prices of inorganic fertilisers in developed countries compared to the costs of animal waste use also militate against their re-use on land. The OECD (Anon., 1986) has examined the problems of animal waste disposal and proposed solutions based upon national 'Codes of Good Agricultural Practice' and better advisory provision to farmers, together with altered pricing policies for fertilisers (including consideration of the 'polluter pays' principle), and the development of new technologies to improve the storage and composition of waste slurries.

Catchment management plans for nutrient removal, from the OECD on a trans-national scale down to basin authorities on a regional scale (such as the Great Lakes which receive approximately one quarter of their phosphorus loads from diffuse agricultural sources (Berg, 1979)) recognise that substantial nutrient reduction is possible, but only with a carefully planned and coordinated strategy at the level of individual farm holdings, incorporating education, soil conservation, careful and more scientific fertiliser application, and better management of animal wastes.

7.2.2 Management of point source discharges

The majority of point source discharges are human or industrial sewage treatment plants, together with larger agricultural units. The sewage treatment process is primarily designed to reduce gross pollution of watercourses by organic matter, which results in the oxidation of nutrients to their biologically available inorganic forms. Technologies to remove nutrients are relatively new and are added onto existing sewage treatment processes at a final stage. This is sometimes referred to as 'tertiary treatment' because it follows the normal two-stage treatment of sedimentation followed by biological treatment. There are many other forms of tertiary treatment however, to upgrade effluents by the further removal of BOD (Biological Oxygen Demand) and suspended solids, so nutrient removal technologies are better referred to under the general term of 'advanced wastewater treatment' (AWT).

Nitrogen removal technologies

Among the commonest AWT technologies are methods for removal of ammonia. This is toxic to fish in addition to being a nutrient in the ionised form, and so ammonia removal from effluents discharging to sensitive watercourses may often be instigated for reasons other than eutrophication control. Most technologies initially convert ammonia to oxidised nitrogen, so nitrogen removal for eutrophication control is usually a two-step process: nitrification (ammonia → nitrate) and denitrification (nitrate → nitrogen gas).

The least expensive nitrogen removal methods are generally biological processes which utilise the natural microbial transformations of the nitrogen cycle (Chapter 2). Nitrification occurs under aerobic conditions (at least 2 mg/l oxygen concentration) which can be maintained by passing the secondary-treated effluent through processes physically similar to that stage – aerated granular beds (gravel or artificial media), activated sludge tanks or rotating discs (Gaunlett, 1980; Hammer, 1986). The processes are temperature- and pH-dependent, with an optimum temperature range of 8–30°C and pH range of 7.5–8.5 (Hawkes, 1983). Biological denitrification can occur in similar physical plant, but requires conditions of lower oxygen concentration and the addition of a carbon source to promote microbial activity. The least expensive and most suitable is

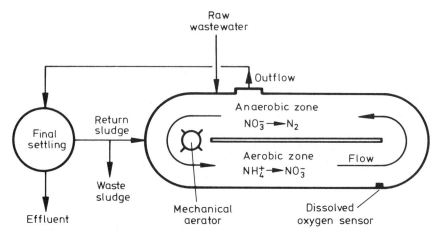

Figure 7.1 Principles of the 'Carousel' aerator for nitrogen removal. Modified from Hammer (1986), with permission.

methanol (McCarty and Beck, 1969):

Combined plant exists where nitrification and denitrification occur in different zones controlled by oxygenation. The 'Carousel' system for example is a vertical-sided, deep, oval tank, with a retention time of around 24 hours. Aeration and circulation is maintained by a mechanical agitator; nitrification progressively gives way to denitrification as circulation proceeds (Fig. 7.1). Total nitrogen removal can be in the order of 50–70% (Hammer, 1986). An alternative is the maintenance of smaller units in series, such as rotating discs. These rotate vertically in an effluent tank, with 50% of the disc submerged at any one time. The disc is of corrugated non-biologically degradable plastic on which the microbial film develops. Exposure to air provides the film with more oxidised conditions for nitrification; immersion provides more oxygen-depleted conditions for denitrification (Antonie, 1978).

Physical and chemical processes exist for ammonia removal, although they are generally more expensive than microbiological ones. Chemical stripping by the addition of lime to effluent which raises its pH above 11, followed by its passage down an aerated tower can effect removal rates of up to 90%. Subsequent adjustment of the pH below 9 is necessary with carbon dioxide (Hammer, 1986). Temperature affects the solubility of ammonia (which is greater, and hence less easily removed, at lower temperatures) and may also result in ice formation in the tower.

Other problems include the formation of calcium carbonate scale or sludge. Less expensive is 'breakpoint chlorination' (which may also be used in potable water treatment processes) whereby the addition of chlorine gas results in the formation of gaseous nitrogen and nitrogen oxides. Drawbacks include the formation of nitrogen trichlorides and the passage of free chlorine through the process which would inhibit subsequent biological denitrification steps.

Ion exchange processes, using zeolite resins at pH of around 6.5 can effect up to 99% ammonia removal as well as substantial nitrate reductions. These are more expensive processes because of capital cost, resin regeneration costs and waste disposal problems. They are more likely to be developed for future treatment of raw ground or river water sources for drinking water supply in the UK (Wilkinson and Greene, 1982) although they replace other chemical wastewater treatment methods in the US during low-temperature periods (Bayley, 1971).

Phosphorus removal technologies

Phosphorus removal is usually effected by chemical methods, although physical and biological methods are also used. Chemical precipitation using iron or aluminium coagulants can take place at any step of the two-stage sewage treatment process: prior to primary sedimentation, to the aeration tank during an activated sludge stage, or following biological treatment. Depending upon the chemical added, ferric or aluminium phosphate is precipitated together with polyphosphates and organic phosphates removed by entrapment or adsorption onto floc particles. Waste pickle liquor from the steel industry (a mixture of ferrous sulphate and chloride) may also be used; it is inexpensive but requires pH adjustment with sodium hydroxide or lime (Hammer, 1986). Precipitation with lime (producing calcium hydroxyapatite) has also been used; this produces larger sludge volumes because of the simultaneous production of calcium carbonate, but the sludge has better thickening and stabilisation properties. Removal of total phosphorus in excess of 90% is achieved (Mekerson, 1973; Kerrison *et al.*, 1989), and treatment may have additional benefits such as the more efficient precipitation of suspended organic solids. Costs are approximately 10–30% in addition to the costs of treatment to the secondary stage. The relative merits of different processes and chemicals have been reviewed by Balmér and Hultman (1988) from the widespread experience of their use in several mainland European countries. Effluent concentrations below 1 mg P/l are readily achieved and lower concentrations (to 0.2 mg/l) can be reached by addition of a filtration step which removes particulate phosphorus at the end of the process.

Biological removal processes (additional to those of conventional treatment which remove 20–30% of incoming phosphorus) are, like the biological nitrogen removal methods, based upon the same principles as secondary biological treatment for BOD removal. They rely on the phenomenon of 'luxury uptake' of phosphorus by certain microorganisms which occurs under anaerobic conditions and is then used for subsequent growth under aerobic conditions (Eckenfelder, 1969). Two activated sludge tanks in series, each with a contact time of 2–5 hours, the first anaerobic and the second aerobic, provide the simplest process. This can be modified by the addition of an anoxic tank which facilitates nitrogen removal (Balmér and Hultman, 1988), or by making the phosphorus removal a side-stream process which gives greater control and flexibility (Tetreault *et al.*, 1986).

Biological phosphorus removal processes are more recent technologies than chemical removal, and are principally being developed in the US and South Africa. In tropical and sub-tropical regions they may be supplemented or replaced by methods involving removal by algae or higher plants which are subsequently harvested. Algae have the disadvantage of contributing to suspended solids and BOD themselves, but plants such as the water hyacinth, *Eichornia crassipes*, a floating plant originally native to South America but now widely dispersed in Africa, America and Asia, can achieve removal rates of 90% for both nitrogen and phosphorus and provide a low-technology alternative for nutrient removal as long as harvesting and disposal can be accommodated.

A recent development of biological tertiary treatment in the northern hemisphere has been the incorporation of units of emergent plants at sewage treatment works, mimicking the nutrient retention of natural wetlands. This 'root zone treatment' is usually based upon the common reed, *Phragmites australis* in Europe. The principle of operation is the microbiological activity associated with the rhizomes of the plant; aeration in the rhizosphere being provided by oxygen transport down the plant tissues. A complex of aerobic and anaerobic reactions occur similar to those described for the more technological kinds of biological removal processes above. Construction of the basin and the nature of the substrate in which the reeds are grown is important, but once established (after about three years), maintenance costs are low and effluent quality high

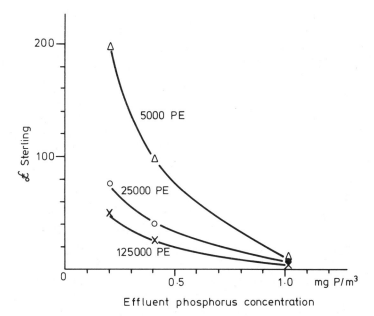

Marginal cost
£ Sterling / kg P removed, 1986 prices

Figure 7.2 Marginal costs of phosphorus removal for different sized treatment works and different effluent standards, costs converted from Swedish K to £ sterling at 1986 prices. Modified from Balmér and Hultman (1988), with permission.

Table 7.1 Additional per capita costs for phosphorus removal after primary and biological treatment for different effluent phosphorus concentrations (PE = population equivalent). From Balmér and Hultman (1988), costs converted to £ sterling at 1986 prices

Population equivalent of treatment works	Effluent total phosphorus concentration required		
	1 mg/l	0.4 mg/l	0.2 mg/l
5,000	5	16	23
25,000	4	9	11
125,000	3	5	7

(Gersberg *et al.*, 1983). Over 400 of these systems are now in operation or being tested in Europe.

Biological phosphorus removal methods are relatively recent and their cost-effectiveness has not yet been evaluated. Chemical methods can more easily be assessed and an example of costs for total and per kilogram P removal, for Sweden in 1986 (Balmér and Hultman, 1988), is given in Fig. 7.2 and Table 7.1. It is clear from the marginal costs that the effluent phosphorus concentration required will be determined on economic grounds in relation to the improvements needed in the receiving watercourse or water body. This will be different in every catchment, emphasising the need for individual catchment decisions for phosphorus effluent concentrations below 1 mg/l.

Political controls

The development and use of AWT processes for nutrient removal has come about as a result of political decisions, made after public concern over eutrophication was aroused and expressed. These decisions have set effluent standards for phosphorus or nitrogen in much the same way as standards were set for BOD and suspended solids earlier this century. The two other areas of political control are the regulation of phosphate content of detergents and the more recent development of controls over fertiliser nitrogen addition in sensitive areas, referred to earlier in this chapter.

Detergent phosphorus concentrations were originally about 10–12% by weight in the 1960s and this contributed about half the per capita load of phosphorus to domestic sewage works (Chapter 2). Many countries have reduced this contribution by legal or voluntary controls to half or less. Voluntary controls by detergent manufacturers have reduced phosphorus percentage content to 7.5% in Sweden, 5% in the Netherlands (Sas, 1989) and 4.5% in the US (Porcella, 1985). Legal controls have reduced this composition even further; to 2% in Italy, 0.5% in the State of Wisconsin, USA, and 0% in Switzerland (Sas, 1989; Pallesen *et al.*, 1985).

In some countries, controls have been catchment-based. In the basin of Lake Mjøsa the promotion of phosphate-free detergents reduced the annual loading on the lake over six years by about a third, reducing the incidence and magnitude of cyanobacterial blooms (Cullen and Fosberg, 1988). The Canadian government reduced detergent phosphorus content to 8.5% in 1970 and 2.2% in 1972; with combined US/Canadian reduction to 0.5% in 1977 (Slater and Bangay, 1979; Cullen and Fosberg, 1988). Phosphorus

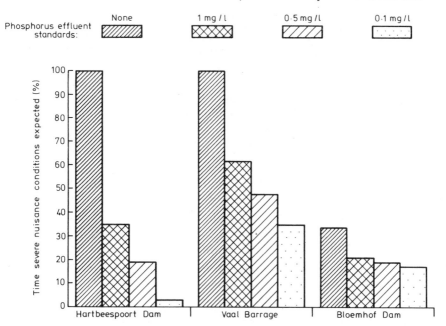

Figure 7.3 Predicted impact of phosphorus control standards on the percentage of time that nuisance algal blooms occurred in three South African reservoirs. Modified from Grobler and Silberbauer (1985a), with permission.

reduction in detergents has largely been achieved by their substitution by other compounds such as citrate and nitriloacetic acid (NTA) (Hamilton, 1972), without reduction of their washing efficiency or unforeseen environmental effects (Fosberg, 1980).

The overall effectiveness of detergent phosphorus reduction on its own has been variable. Measured reductions of around a quarter of pre-ban inflowing phosphorus have been recorded in studies in the US states of Michigan, Wisconsin and New York, but without significant effects on the quality of receiving lakes (Maki, 1984; Lee and Jones, 1986). Prediction of the effects of detergent phosphorus removal from the catchment of three eutrophic South African reservoirs suggested that even 100% removal would have no impact on algal nuisance conditions. In most cases it appears that detergent phosphorus removal has to be combined with wastewater treatment to reduce phosphorus loads to lakes sufficiently for improved conditions to be observed (Grobler and Silberbauer, 1985a) (Fig. 7.3).

Legal controls on effluent phosphorus concentrations have been implemented in some European countries and US states. Sweden has effectively imposed effluent standards of 0.5 mg P/l from urban sources since 1972, Switzerland 0.8 mg P/l since 1985. Germany required all sewage plants serving more than 50,000 inhabitants to achieve a 2 mg P/l standard from 1988 and the Netherlands will impose a similar policy with stricter limits on about one-third of effluents from 1992 (Sas, 1989). In South Africa a national standard of 1 mg P/l is being progressively implemented, and in Australia limits of 0.2 mg P/l are imposed for Canberra and are being considered for Sydney (Cullen and Fosberg, 1988). At the beginning of 1990 a Directive of the European Commission was published requiring that member states adopt a standard of 1 mg/l P on municipal wastewater treatment plants discharging to 'sensitive areas' by 1998. Sensitive areas are defined as 'natural freshwater lakes, other freshwater bodies, estuaries, coastal waters and seas which are found to be eutrophic or which in a short time may become eutrophic if protective action is not taken' and 'areas of high ecological quality due to the flora and fauna present and other areas which are important from a scenic or nature protection point of view'.

7.2.3 Diversion of inputs

This technique is not widely available, depending upon the ability to route effluents away from the lake catchment into an adjacent one of lesser importance, or the sea. Two case studies of effective effluent diversion are well known: Lake Washington in the US where effluents were diverted to the sea and successful alleviation of the symptoms of eutrophication were achieved (Edmonson and Lehman, 1981) and Lake Norrviken, Sweden, where effluents were diverted to a new sewage treatment plant on the coast (Ahlgren, 1972, 1978) but lake improvement did not occur because of sediment phosphorus release (see below). Eight others in the US, Sweden and Austria were summarised by Cullen and Fosberg (1988).

7.2.4 Upstream sedimentation

In certain circumstances, sedimentation of phosphorus can be achieved upstream of the inflows to a lake if an impoundment exists or can be created. This happens naturally in chains or 'cascades' of reservoirs on a river system. An interesting example was provided by the Jesenice reservoir in Czechoslovakia (Fiala and Vasata, 1982) where a new reservoir created in the early 1960s was divided into two parts by a railway embankment. The 'pre-reservoir' upstream of the embankment functioned as a settling pond with a retention time of five days, eliminating about 70% of total P and 40% of N from the inflow, as well as providing important reductions in coliform bacteria.

7.3 EVALUATION OF NUTRIENT CONTROL MEASURES

A large number of nutrient control programmes for lakes, mainly in the northern hemisphere, have been implemented and the effectiveness of control programmes can be evaluated. Cullen and Fosberg (1988) list over 40 such lakes, and divide their responses into the following classes in order to facilitate interpretation of the varied results:

Type 1 response
   Reduction in phosphorus and chlorophyll concentration sufficient to change the trophic state of the water, based upon the trophic state categories of Fosberg and Ryding (1980) (see Chapter 6).
Type 2 response
   Reduction in phosphorus and chlorophyll insufficient to change the trophic category, but sufficient to cause noticeable improvement in trophic conditions.
Type 3 response
   Little or no obvious improvement, such as reduction in lake phosphorus but little or no reduction in chlorophyll.

Approximately 40% of lakes reviewed showed a type 1 response, 20% type 2 and 40% type 3. Reasons for the failure of nearly half the lakes to show a response are varied, and depend upon the characteristics of the individual lake and its catchment. In some cases point source reductions were not enough because of diffuse sources. In others the water renewal time was insufficient for recovery to be observed during the duration of the reported study. In some lakes phosphorus removal alone had no effect because it was not the limiting factor, which was usually nitrogen. In many however, internal loading from sediments prevented immediate recovery. Interpretation of failure was difficult in some cases where diversion removed both nitrogen and phosphorus, because nitrate removal may enhance internal phosphorus loading, and low nitrate concentrations may favour cyanobacteria development at the expense of other algal species.

Sas (1989) reported a more critical cooperative study of the restoration by phosphorus loading reductions of 18 eutrophic lakes in Western Europe. The participants concluded that deep lakes (with most of the epilimnion not in contact with the sediments) behaved differently from shallow lakes. They divided the analysis of responses into two categories, the response of lake phosphorus concentration, and the response of the algal community (by chlorophyll, cell volume, transparency and species changes).

In the shallow lakes a strong net annual release of phosphorus from sediments occurred after restoration, delaying the reduction of lake phosphorus concentrations by up to five years. No such annual release occurred in deep lakes. Both kinds of lakes showed seasonal sediment releases, which are of considerable importance in shallow lakes because release is directly into the epilimnion at a time when algal growth response can be rapid. The clearest guide to sediment behaviour after restoration appeared to be the sediment phosphorus concentration in the upper 15 cm; above 1 mg P/g dry sediment several years of phosphorus release occurred. The influence of oxygen conditions in the water above the sediment surface did not appear to be so important as phosphorus content. Deep lakes, even those with similar pre-restoration loadings as shallow lakes, had lower sediment phosphorus concentrations; this was attributed to the additional effect of mineral particles and the relative lack of disturbance of sediment surface.

The responses of the algal communities were explained by a four-stage conceptual model, although not all lakes passed through all stages.

Stage 1

This occurred in lakes subjected to high phosphorus loadings, where no effect on the phytoplankton was observed following reduction of lake phosphorus concentration. Direct phosphorus limitation on algal growth was not apparent until average soluble phosphorus concentrations in the epilimnion fell below about 10 µg P/l. Nitrogen control of biomass occurred in some lakes, and it was tentatively concluded that the threshold for biomass reduction was 100 µg N/l.

Stage 2

This was a behavioural response of phytoplankton, confined to deeper lakes. Here phosphorus limitation resulted in greater dispersal of motile species within the water column below the epilimnion, with concomitant increase in transparency. Shallow lakes did not show this response.

Stage 3

This was a decline in mean phytoplankton biomass, close to that predicted by the Vollenweider/OECD model. There was some evidence that this was slower in the shallow lakes than the deep ones. Transparency responses were more varied.

Stage 4

The final stage was a change in algal species composition. The pre-restoration dominance of cyanobacteria was shown to be substantially reversible once a threshold of phosphorus concentrations of 50–100 µg P/l in shallow lakes and 10–20 µg P/l in deep lakes was passed. Shallow lakes previously dominated by *Oscillatoria* species showed more complex responses, because *Oscillatoria* tended to disappear at higher phosphorus concentrations and then be replaced by other cyanobacteria forming surface scums. Circumstantial evidence from the studies suggested that the most important additional factor determining cyanobacteria abundance was light penetration. Cyanobacteria maintain populations at low light intensities yet are relatively easily light-inhibited, and this explained their individual responses in the lakes.

7.4 CONTROL OF NUTRIENT CONCENTRATIONS WITHIN LAKES

Alternatives or supplements to the removal of nutrients from catchment sources are their removal from within the lake. A number of methods exist, although none have yet been widely used because of cost or technical limitations.

7.4.1 Lake flushing and hypolimnetic outflow

Lake flushing requires a readily accessible source of low-nutrient water. Two lakes in the United States have been successfully treated in this way, but the places where such techniques can be used are limited to those close to such sources. Green Lake, Seattle, USA was suitable for such treatment because the city possessed an abundant supply of high quality water (which only required disinfection for treatment). Dilution from this source gave the lake a renewal time of four months and resulted in a reduction of total phosphorus concentration from 60 to 20 µg/l, chlorophyll from 50 to 10 µg/l and a transparency increase from 1 to 4 m in three years (Oglesby, 1969). Moses Lake, Washington, was flushed periodically during spring and summer using water from the nearby Columbia river irrigation system producing 40–60% improvement in chlorophyll, phosphorus and transparency (Welch and Patmont, 1980).

Hypolimnetic outflow is an alternative method of removing nutrients, usually from deeper eutrophic lakes, when stratification sets in during the summer. Its drawbacks are the possible effects of low temperature, low oxygen and high nutrient water on the receiving stream. The technology is basically simple – a siphon from the deepest part leading to the outflow – and the method has been used successfully in Switzerland and several other European locations (Welch and Patmont, 1980). Lake Bled, Yugoslavia, experienced improved conditions following the introduction of hypolimnetic outflow after inflow flushing from the nearby River Radovna failed to produce any improvement (Vrhovsek et al., 1985).

7.4.2 Sediment removal

Sediment removal is an expensive technique, both in the pumping equipment costs and the sludge transport and disposal. It has been used as a last resort on shallow lakes with high public value where other techniques of nutrient limitation have failed due to the magnitude of sediment phosphorus release. Lake Trummen, a shallow lake (maximum depth 2.4 m) in Sweden, was highly eutrophic by the late 1950s as a result of effluents received over about thirty years. Effluent diversion after 1958 failed to effect any recovery due to high internal loadings. Suction dredging was employed during 1970–71, removing the upper half metre of sediment and discharging it to adjacent settling ponds. An immediate improvement occurred in phosphorus and algal biomass reduction (Andersson et al., 1973; Bengtsson and Gelin, 1975) (Fig. 7.4). The cost, however, was about £250,000, equivalent at 1970 prices for 100 ha, and would probably be at least five times higher now. Peterson (1979) reviewed the techniques and problems of this restoration method in America, and concluded that cost and disposal of sediment were the main limitations. In the Norfolk

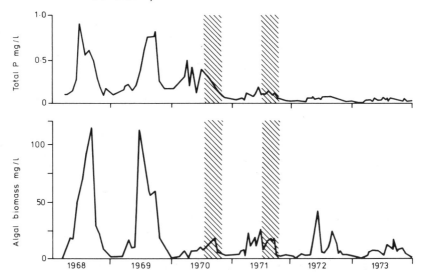

Figure 7.4 The effects of sediment removal upon phosphorus and algal biomass at Lake Trummen, Sweden (hatched areas are periods of sediment removal). Modified from Bengtsson and Gelin (1975), with permission.

Broads, England, sediment pumping is being developed as a successful though expensive restoration technique, where nutrient removal alone has not been successful because of sediment phosphorus release compounded by the stirring action of recreational boat traffic (Anon., 1987).

7.4.3 Nutrient inactivation by precipitation or bottom sealing

Nutrient precipitation has been experimented with in small lakes, using the same principles as those employed in AWT methods of effluent treatment: addition of iron, aluminium or similar salts. Most studies have reported initial success in reducing phosphorus and chlorophyll, but limited persistence of the effects in the absence of renewed additions. Welch (1983) reported on the success of several American trials. Foy (1985) showed in a five year study of a lake treatment in Northern Ireland that the first year after addition of ferric aluminium sulphate resulted in hypolimnetic phosphorus release reduced by 92% compared to the pre-treatment year, but two years later the situation had returned to pre-treatment levels. Reduction in winter lake phosphorus content occurred each year due to hypolimnetic mixing of Fe and P at overturn and subsequent precipitation. Epilimnion phosphorus concentrations however, were little affected by these changes, probably because of release from epilimnetic sediments between April and June. One of the problems of sediment inactivation of phosphorus in this way is the effect of aquatic macrophytes which take up phosphorus from below the sediment surface and release it through subsequent senescence into the epilimnion. Macrophyte growth may additionally be enhanced by release from light limitation as a consequence of phosphorus restriction of algal growth. A small drinking water supply reservoir in south-east England, Foxcote, experienced prolific growths of rooted plants and filamentous algae three years after ferric sulphate dosing successfully reduced the planktonic algae which had caused water treatment problems (Young *et al.*, 1988).

Sediment immobilisation by bottom sealing with plastic sheeting has been tried experimentally but is expensive and limited by technical problems such as the release of gas from the sediments.

7.4.4 Biomass harvesting

Removal of biomass with its accumulated phosphorus is a superficially attractive technique, but is really only feasible for aquatic plants and, to a limited extent, fish. The proportions of lake phosphorus potentially removable in this way depends upon the proportion of littoral zone in any lake colonised by plants, but may be up to 60% of input in favourable circumstances (Welch, 1983). Macrophytes may also remove much of their phosphorus from an otherwise immobile source below the sediment surface. Plant harvesting is practised widely in the United States but mainly as a management tool for reducing nuisance growths in recreational lakes and avoiding fishkills as a result of periods of low oxygen concentrations rather than for nutrient removal. Only in certain circumstances where very dense growths occur, such as those of water hyacinth, is nutrient control possible (Burton *et al.*, 1979). Harvesting of fish and algae have also been considered, but even in highly eutrophic tropical lakes (Thornton *et al.*, 1986), their effectiveness as removers of nutrients is too small to be practical (Bull and Mackay, 1976).

7.5 MANAGEMENT OF LAKES WITHOUT NUTRIENT REDUCTION

Three methods exist for managing nutrient rich lakes without altering their nutrient state (although they are increasingly used in conjunction with nutrient reduction). A physical method, artificial mixing or aeration, has been widely used all over the world for thirty years or more. A biological method, manipulation of lake communities is more recent in its application although the ecological theory dates back twenty five years. The third method is treatment of algal and/or rooted plant growth by herbicidal chemicals.

7.5.1 Artificial aeration and maintenance of mixed conditions

One of the most serious effects of eutrophication is that the decay processes of a lake, acting upon high quantities of detritus from the production of the epilimnion, cause partial or complete deoxygenation in the hypolimnion (Chapter 1). The reducing conditions which then occur in the hypolimnion lead to the re-solution of iron and manganese, the production of ammonia and sulphides, and the release of phosphates and silicates. The iron and manganese together with low oxygen concentration cause water treatment problems, and the nutrients ultimately support further algal growth (Fig. 7.5, Table 7.2).

Artificial destratification to maintain the water body in an isothermal condition, or artificial aeration of the hypolimnion without destroying the stratification, are methods which were first developed to prevent chemical changes from occurring. In the last ten years these techniques have also been investigated as a means of controlling algal populations. The design and capacities of the equipment to be used depend upon the lake depth, the volume of water to be re-aerated or overturned, and its temperature and oxygen conditions. They thus have to be calculated for each individual water body, most easily using guidelines synthesised from the experience of operation elsewhere (Lorensen and Fast, 1977; Tolland, 1977; Ashley, 1985). Both kinds of method may be operated continuously from the period immediately before stratification is expected to set in, or intermittently to re-aerate or to break down stratification as necessary. A monitoring programme to control the operation and its running costs is usually necessary.

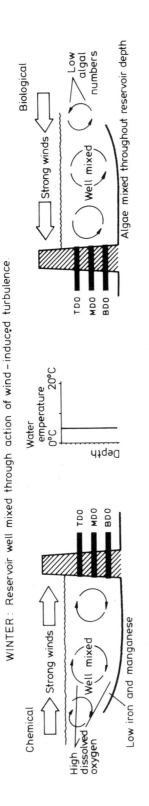

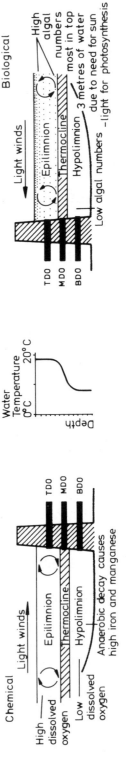

Figure 7.5 The consequences for water treatment of low oxygen concentrations in a reservoir hypolimnion. Modified from Brierley (1984), with permission.

Table 7.2 Management problems of eutrophic reservoirs during stratification and deoxygenation of the hypolimnion. From Brierley (1985)

Choice of reservoir draw-off	*Treatment problems*
Top draw-off High dissolved oxygen, low ammonia, iron and manganese but high algae	High algal numbers (a) raise the pH of the raw water which creates problems with coagulants; (b) reduce filter runs; (c) increase chemical dosing costs; (d) cause taste and odours in final water; (e) and can break through into into distribution system providing food for animals in the water mains
Middle draw-off Adequate dissolved oxygen, low ammonia iron manganese but can have high algal numbers and sedimenting algal cells	
Bottom draw-off Low algal numbers but low dissolved oxygen; high iron, manganese and ammonia	High ammonia – increases chlorine dose required for sterilisation High iron and manganese – require extra chemical treatment to prevent discolouration

Artificial aeration

The principle of artificial aeration of the hypolimnion is to prevent anaerobic conditions developing whilst maintaining the lower temperatures (an advantage in water treatment and the maintenance of a cold-water fishery) and the accumulation of nutrients (to keep them unavailable for further algal growth) which naturally occur. The equipment draws water into the bottom of a tube extending from just above the sediment to the water surface by the introduction of compressed air (or sometimes oxygen) at the bottom through a diffuser which creates fine bubbles. Outlet tubes at the top of the hypolimnion release the oxygenated water whilst the air continues to the top of the tube (Fig. 7.6).

This method is best suited to eutrophic lakes and reservoirs where regular, stable summer stratification occurs. It has been successfully used for reoxygenation of the hypolimnion and the creation of an oxidised mud layer of 2 mm thickness in Lake Jarlasjon in Sweden (Bengtsson and Gelin, 1975), and since 1966 in the Wahnbach reservoir of Germany (Bernhardt, 1977). It has generally been

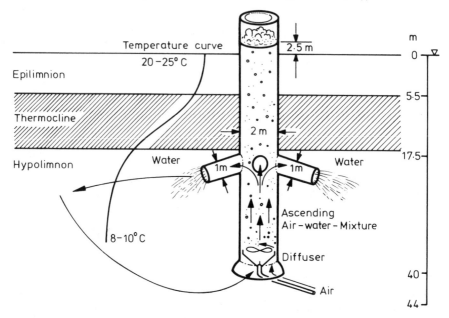

Figure 7.6 A hypolimnion aeration device. Modified from Tolland (1977), with permission.

successfully applied in the US in a variety of variations on the basic design (Fast, 1979).

The biological effects of hypolimnetic aeration are varied. There is little direct effect upon phytoplankton, because the epilimnion is undisturbed. Zooplankton populations may be enhanced, because they are more able to migrate vertically downwards into an oxygenated hypolimnion by day and avoid predation. In Hemlock lake, Michigan, this allowed the development of a population of large-bodied *Daphnia pulex* which was virtually absent before areation (Fast, 1979). In theory this heavier grazing pressure would reduce algal biomass although evidence is inconclusive. Fish diversity and biomass may be enhanced, because a eutrophic lake with an aerated hypolimnion could support both warm water and cold water fishes when formerly cold water species would not survive in a low-oxygen hypolimnion.

Maintenance of mixed conditions (artificial destratification)
Artificial destratification may be achieved in three ways:

1. Introduction of compressed air through a perforated pipe or diffuser, so that the fine bubbles released entrap bottom water and circulate it to the surface.
2. The mechanical pumping of water from the hypolimnion to the epilimnion.
3. The 'jetting' of inflow water under pressure into a reservoir if it has pumped inflows.

The first method is the commonest. Commercial designs such as 'Helixor' tubes exist (Fig. 7.7), which usually need to be installed in a reservoir at the construction stage or in a lake during temporary drainage if this is possible, at the deepest point. These create a rising plume of bubbles. A simpler alternative is the sinking of a pipe of flexible material with regular small perforations along its length along the deepest axis of the lake connected to a compressor or

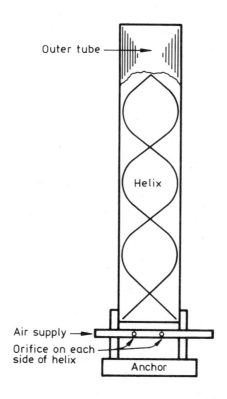

Figure 7.7 A 'Helixor' tube for artificial overturn of reservoirs. Modified from Tolland (1977), with permission.

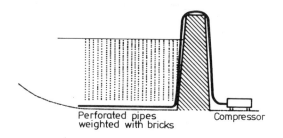

Perforated pipes
weighted with bricks   Compressor

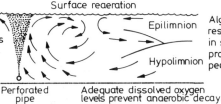

Surface reaeration

Bubbles of air entrain
water as they rise –
water from lower levels
carried to surface.
Local destratification
followed by general
destratification

Epilimnion

Hypolimnion

Algae mixed throughout
reservoir-decreases time
in sunlight -should reduce
productivity thus reducing
peak algal numbers

Perforated
pipe

Adequate dissolved oxygen
levels prevent anaerobic decay

Figure 7.8 Perforated pipe system of reservoir overturn and its principle of operation. Modified from Brierley (1984), with permission.

other source of compressed air. This creates a 'curtain' of bubbles. Both techniques destratify the immediate region of their operation and then progressively draw in water from further away until the body is fully mixed (Fig. 7.8). The time which this takes depends upon the power of the equipment and the climatic conditions pertaining at the time – hot, calm conditions will act to re-establish stratification while cool windy conditions will aid destratification.

Providing that the power of the equipment is adequate, artificial destratification is effective in maintaining a fully oxygenated water column with the suppression of release of iron, manganese and nutrients (Fig. 7.5). It is inexpensive, costing about 5% of the total treatment costs of a waterworks (Brierley, 1984) (Table 7.3).

Pumping and inflow jetting (Steel, 1972) achieve the same ends – a fully mixed water column throughout the summer season when stratification would otherwise set in. The latter technique can usually only be operated continuously.

Table 7.3 Costs of destratification at Staunton Harold reservoir, English midlands, compared with chemical treatment costs at Melbourne treatment works serving two reservoirs. From Brierley (1984), 1979 prices

Destratification costs (£ sterling)		Chemical treatment costs (£ sterling)
Electric compressor, pipework, installation	4,000	104, 200
Running costs per annum	100	

The effects of artificial mixing upon phytoplankton populations are apparently conflicting, however (Fosberg, and Shapiro, 1980). If the main effect is to recycle nutrients which would otherwise be trapped in the hypolimnion back into the photic zone of the lake, then algal growth will be enhanced. On the other hand, if algal cells which would otherwise be confined to the narrow photic zone are circulated throughout the depth of the water column, and assuming that the mixed depth under these circumstances exceeds the euphotic depth, they experience a lower overall light regime and their photosynthesis and growth will be inhibited. Changing the light regime or the nutrient regime in a lake can also be expected to change the competitive advantages of one species over another, and thus alter the pattern of algal succession. In the range of studies reported both increases (e.g. Fast, 1981) and decreases (e.g. Cowell *et al.*, 1987) of biomass and/or primary production have been found after mixing.

These apparently conflicting results are explicable on the basis of the capacity of the individual lake's destratification equipment to effectively mix the water column, coupled with the explanation of algal species succession developed by Reynolds and others (Chapter 4), which were confirmed in a series of controlled experiments in limnetic enclosures (Reynolds *et al.*, 1983, 1984). Reynolds and colleagues used periods of intermittent mixing which promoted the growth of spring–autumn diatom species; heavy cells adapted to low light conditions. Subsequent stratification eliminated these mixed-column species and resulted in the growth of summer 'r'-strategists such as small green algae. Alternate periods of mixing and stratification of about 3-weeks' duration prevented either type of algae from achieving their maximum potential biomasses. Slow growing, 'K' selected species slowly increased in abundance over the whole period of the project, but probably reached their peak at a lower biomass and later in the year than they would have done had more stable conditions existed. These experimental results support observations on the operation of intermittent mixing in reservoirs where population developments of individual species are not able to achieve their maximum (Ferguson and Harper, 1982) and where succession can be explained in terms of the Reynolds' matrix of nutrients and column mixing (Fig. 7.9) (Brierley, 1985). These explanations of changes in terms of the reproductive strategy of the taxonomic forms concerned should not be viewed in isolation from explanations which seek the more detailed reasons for shifts; indeed the two approaches go hand-in-hand. Shapiro (1979) has summarised the range of mechanisms for effects of artificial mixing upon the succession of algal taxa and upon biomass (Fig. 7.10), and further research will enable these to be quantified.

Mixing may also alter zooplankton populations. There are conflicting results here too, with most studies reporting an increase in both individual size and total numbers of zooplankton following mixing (Pastorak *et al.*, 1980) but some reporting the opposite (Cowell *et al.*, 1987). The more usual observation, of increase, is explained by the distribution of zooplankton in a greater volume of water, reducing interactions with fish at lower light intensities, and hence reducing predation pressure upon the zooplankton. In shallower hyper-eutrophic lakes (Cowell *et al.*, 1987) this effect would be lessened.

7.5.2 Biological manipulation of communities

Biological manipulation – 'Biomanipulation' – may be defined as the deliberate alteration of one component of the limnnetic food web in order to promote changes in the other components with the aim of alleviating some of the symptoms of eutrophication (mainly high algal biomass and low transparency). It is usually the alteration of piscivorous or planktivorous fish by stocking/removal, but may also include the direct alteration of zooplankton by mixing.

The principles of biological manipulation of communities to alleviate the effects of eutrophication were developed as a result of observations on the effects of predation by fish upon zooplankton. Research in fish ponds in Eastern Europe (Hrbacek *et al.*, 1961) and on the effects of introduced planktivorous fish into lakes (Brooks

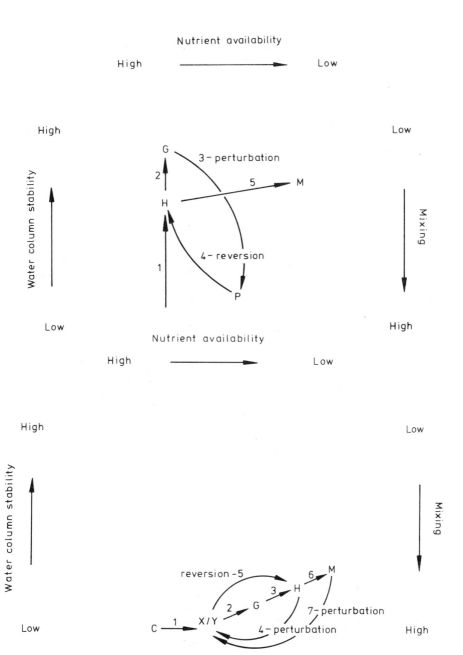

Figure 7.9 Algal succession in an English reservoir over two years of operation of artificial mixing, based upon Reynolds' matrix (see Chapter 4) for intermittent artificial mixing perturbations. Modified from Brierley (1985), with permission.

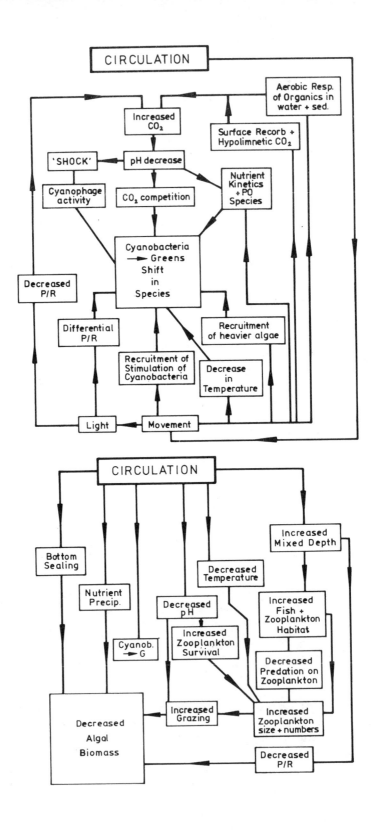

and Dodson, 1965) showed that the body size and abundance of zooplankton was reduced in the presence of fish as a result of size-selective predation, leading to changes in species composition in favour of smaller bodied forms (Fig. 4.23). In the absence of significant fish predation, larger bodied forms which are more efficient grazers of algae (Burns, 1968, 1969) out-compete the smaller species (Hall *et al.*, 1976), although other factors such as invertebrate predation and selection pressures leading to repro-duction at smaller sizes complicate the picture (Lynch, 1977). Grazing zooplankton, notably cladocera, are capable of filtering the entire volume of a eutrophic lake several times over each day at densities achieved in the summer (Porter, 1977), so mechanisms which favour the persistence of large-bodied species such as *Daphnia* – low fish predation pressure – will maintain a standing crop of algae potentially below the carrying capacity defined by the nutrient status of the lake.

The development of these theories into explanations of the extent of zooplankton control of algal biomass in lakes of differing trophic state (Gliwicz, 1975; Nilssen, 1978) has led to the concept of 'top-down' control of plankton communities, in contrast to the 'bottom-up' control by nutrients (Carpenter *et al.*, 1985; McQueen *et al.*, 1986). Studies of whole lakes subject to changing status of fish populations and fisheries management have lent support to the importance of top-down control. In Lake Kinneret, Israel for example, where fish introductions have increased predation pressure on zooplankton and this has lessened grazing pressure on nannoplankton, summer algal blooms and deterioration of water quality occur (Gophen, 1984).

Techniques which maintain the abundance of large-bodied zooplankton can thus result in low algal standing crop and clear water conditions. Control of the fish community by selective removal and re-stocking of lakes offers such a mechanism (Shapiro and Wright, 1984; Van Donk *et al.*, 1989) and is generally successful. Experiments in ponds or limnetic enclosures have also confirmed its importance (Andersson *et al.*, 1978; Lynch and Shapiro, 1981). The actual mechanisms for the maintenance of grazing zooplankton in shallow lakes are more complicated, and involve the littoral macrophyte community as a provider of shelter

Figure 7.10 Diagrammatic representation of the biological effects of artificial mixing, upon algal species changes (upper) and biomass (lower). Modified from Shapiro (1979), with permission.

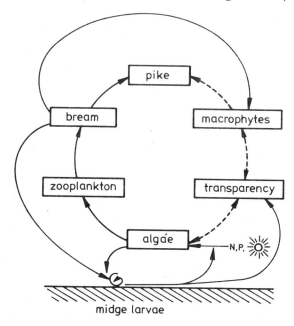

midge larvae

Figure 7.11 Interactions between fish, plankton and plants in shallow lakes. Modified from Hosper (1989), with permission.

for the zooplankton from planktivore predation, and shelter for piscivores such as pike (*Esox lucius*) hunting planktivores (Timms and Moss, 1984; Hosper, 1989). In the absence of macrophytes, pike progressively disappear, planktivore species such as roach (*Rutilus rutilus*) and bream (*Abramis brama*) increase and predation pressure on zooplankton reduces their size and abundance. Released from grazing control, phytoplankton grow until nutrient- or light-limited, and nutrient concentration is often further enhanced by the disturbance of sediment by bream seeking alternative benthic invertebrate food (Fig. 7.11).

In deeper lakes there may be additional benefits of biomanipulation. Migrating *Daphnia* may also excrete substantial quantities of phosphorus into the hypolimnion (Wright and Shapiro, 1984). Macrophyte refuges for predatory fish and zooplankton are of less importance but deeper, darker water may offer a similar refuge.

Experiences of biomanipulation have been accumulating for the past ten years, and syntheses of the results obtained have been attempted in order to better understand the processes and predict the results. One of the negative effects has been the increase in

relative abundance of inedible phytoplankton – usually cyanobacteria – after biomanipulation (Benndorf, 1987). This runs
contrary to expectation (Nilssen, 1978) because cyanobacteria
should be inhibited if efficient grazing maintains high light
penetration and high nutrient concentration with low algal biomass.
It may be that the resistance of cyanobacteria to grazing allows
them to increase slowly in numbers even under unfavourable
physical and chemical conditions. McQueen *et al.* (1986) have
suggested, from empirical analysis of food chain relationships in the
literature and from enclosure experiments, that in eutrophic lakes
top-down effects are strong for piscivore → zooplankton, weaker
for planktivore → zooplankton and have little impact for
zooplankton → phytoplankton. This is partially supported by the
synthesis of available field data (McQueen, 1990), which shows
that 100% of thirty six studies found significant piscivore →
planktivore interaction, 88% significant planktivore → zooplankton
interactions and 64% significant zooplankton → phytoplankton
effects. Both Benndorf (1987) and McQueen (1990) have produced
conceptual models to try to explain the range of effects which occur
as a result of 'top-down' controls meeting 'bottom-up' controls.
McQueen approached the explanations by separating the two
control mechanisms and then attempting a synthesis (Fig. 7.12),
whilst Benndorf compared the combined effects of top-down and
bottom-up in schematic food webs of oligotrophic and eutrophic
lakes with and without biomanipulation (Fig. 7.13).

Both models emphasise the eventual dominance of inedible algae
in the high nutrient state, leading to the conclusion that in the
longer term, biomanipulation alone will not maintain 'oligotrophic'
conditions in the absence of nutrient reduction mechanisms. It is
possible that the additional 'side effects' of biomanipulation, such
as the reduction in bioturbation of sediments if planktivorous
(strictly omnivorous) fish are controlled by piscivores, the interactions of aquatic macrophytes as nutrient and light competitors of
phytoplankton as well as refuges, and the excretion-mediated
removal of phosphorus by migrating zooplankton, might prolong
the beneficial consequences. Longer-term studies are needed to
evaluate these, however.

It is also probable that more detailed understanding of mechanisms as yet little understood or still providing conflicting results will
enable biomanipulation to be used more precisely, probably
through continuous intervention rather than as a once-off tool.
Little-understood mechanisms include negative feed-backs as a

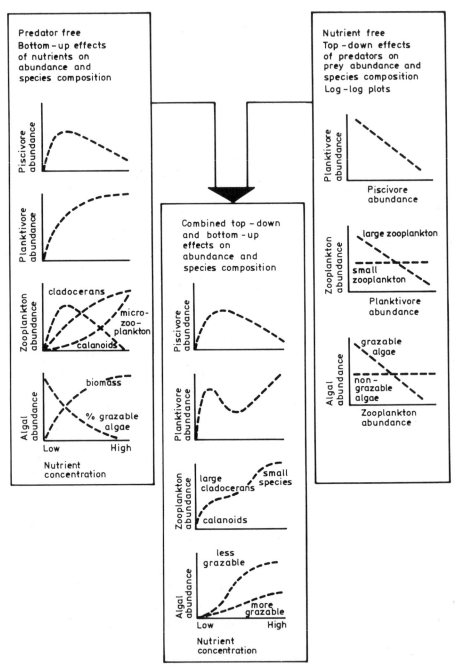

Figure 7.12 Graphical representation of the effects of nutrient concentration on the abundance of organisms (left column), the effects of predation on prey abundance (right column) and the two combined (middle column). Modified from McQueen (1990), with permission.

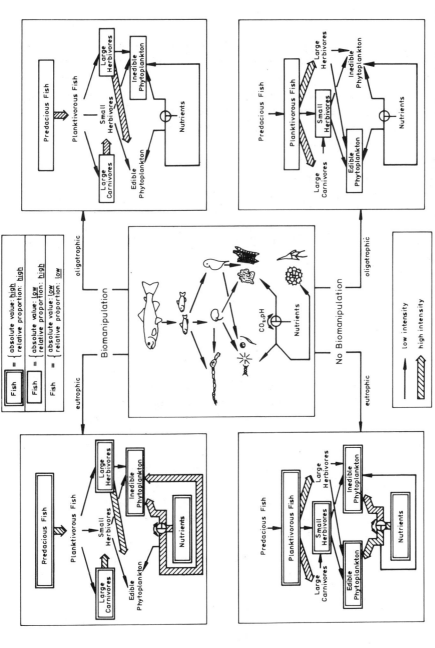

Figure 7.13 Diagrammatic representation of the combined effects of nutrient enrichment and predation with and without biomanipulation. Modified from Benndorf (1987), with permission.

result of zooplankton starvation and the consequences of loss of planktivore biomass through death or weight loss at high density (Threlkeld, 1988). One particularly important area where conflicting results need to be reconciled is the relationship between filter-feeding zooplankton and cyanobacteria (de Bernardi and Giussani, 1990; Gliwicz, 1990), where a variety of factors – such as colony size, toxin production – make cyanobacteria poor food for herbivores. Nevertheless there are examples of zooplankton control of cyanobacteria population growth (van Donk *et al.*, 1990a) as well as examples of lack of control (van Donk *et al.*, 1990b).The differences seem to stem from the relative palatibility of different species of cyanobacteria and the relative efficiency of zooplankton species of different body size. A clearer understanding of other influences on zooplankton, such as possible poisoning by pesticides during the periods of their widespread agricultural use in the 1950s (Stansfield *et al.*, 1989) will also be important. One aspect of biomanipulation which is developing rapidly is the importance of the creation and nature of refuges from predation (Shapiro, 1990b). Seven types were proposed by Shapiro from the literature of biomanipulation:

1. Light refuges of lower light intensity where planktivores cannot effectively see their prey.
2. Low temperature refuges inhabitable by invertebrate herbivores but not their fish predators.
3. Low oxygen refuges habitable by herbivores but not by fish.
4. Macrophytes or artificial physical refuges which obstruct visibility or movement by predators but not herbivores.
5. Open water refuges where the 'visual clutter' of algal colonies may reduce the efficiency of fish predation on herbivores.
6. Behavioural refuges where planktivores in the presence of piscivore populations may restrict their feeding habitats.
7. Predator inefficiency refuges where some planktivorous fish are less efficient than others and may allow persistence of large-bodied herbivore populations.

In tropical bodies biomanipulation has been hardly tried, but experience in the sub-tropics (Florida) suggests that it may be different but simpler (Crisman and Beaver, 1990). Such lakes are characterised by year-round consistency of algal biomass and species composition, with smaller-bodied cladocerans and 'pump-feeding' rather than visual-feeding planktivorous fish. Thus the temperate approach of enhancing large-bodied grazers is not applicable. A greater range of fish species, particularly those which are phytophagous such as the silver carp (*Hypophthalmichthys molitrix*) in Brazil (Starling and Rocha, 1990), have been shown to provide effective phytoplankton control and warrant further investigation.

Biomanipulation will probably become the future method for 'fine-tuning' of lakes after other restorative methods have been applied to the sources of nutrients. The ability of limnologists and lake managers to biomanipulate depends very much upon the relative power of fisheries interests (Hosper and Jagtman, 1990). Some of these may view any removal and alteration of fish stocks with alarm. In the UK for example, some 4–5 million people fish as a recreational activity, returning coarse fish to the water body when caught. The angling clubs which represent these people generally have a very negative attitude towards piscivorous fish unless their members are 'pike specialists'! In Israel too, limnological considerations take second place in fisheries management decisions (Gophen, 1984).

– 8

A case study in restoration: shallow eutrophic lakes in the Norfolk Broads

8.1 INTRODUCTION

The Norfolk Broads, a series of lakes in lowland Eastern England, provide an excellent case study of the effects and the management of eutrophication. The history of the Broads and of the development of eutrophication is reasonably well understood from paleo-limnological studies and historical records. The causes of enrichment bring into focus the problems of modern, highly intensive agriculture and urban sewage disposal. The biological effects are acute because of the value of the area for wildlife conservation, fisheries and recreation, and the consequent economic value of these uses to the local communities. A range of management strategies has been applied, with varying degrees of success, which provide guidance for similar strategies elsewhere. The development of our scientific understanding of the problem, its many interconnections, and the links between understanding and decision-making, offer us lessons in the application of science and technology to the solution of contemporary problems.

8.2 HISTORICAL PERSPECTIVE

The Norfolk Broads consist of a series of very shallow lakes (ranging in size from 1 to 128 hectares and approximately 1 metre water depth) interconnected by rivers and surrounded by marshes and fens found in the eastern part of England (Fig. 8.1). The area is an important wetland in the United Kingdom with three National Nature Reserves – The Bure Marshes, Hickling Broad and Ludham Marshes – together with 24 Sites of Special Scientific Interest and numerous other areas managed for nature conservation. The lakes are man-made, having been formed in the fourteenth and fifteenth

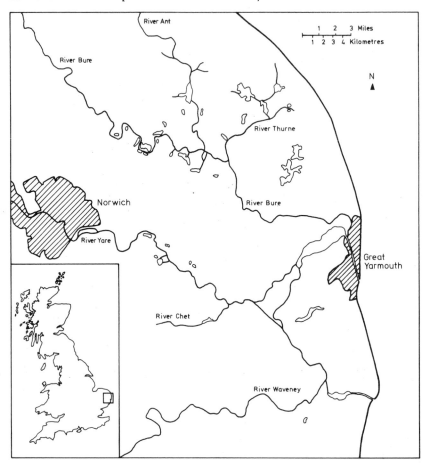

Figure 8.1 Map of the Broads, shallow artificial lakes in the lowland river valleys of Norfolk, Eastern England.

centuries when rising sea levels flooded mediaeval peat diggings (Lambert *et al.*, 1960). Many of the lakes are connected *via* main rivers, notably the Rivers Yare, Bure, Ant and Thurne, forming a substantial area of public navigation, although the majority of the lakes are small and isolated from the navigable river system.

Although we have no detailed information of the ecology of these lakes prior to the 1960s it is clear from naturalists' records that during the early part of the nineteenth century they contained clear water and diverse submerged vegetation dominated by *Chara, Najas* and other plant genera characteristic of naturally rich lowland lakes. Evidence that these communities were widespread

Figure 8.2 The extensive emergent, floating-leaved and submerged plant community of the Norfolk Broads at the turn of the century. Published with permission of Norfolk Library and Information Services.

comes from examinations of sediment cores where remains of *Chara* have been found from numerous broadland sites (Moss, 1979b; Moss, 1980; Moss *et al.*, 1979a). This community was described in Barton Broad around 1903 (Nicholson, 1900) and a very similar flora could be found at Hickling Broad until the 1960s (Phillips, 1963). Around the edge of the lakes there was a gradual transition from the open-water submerged plant communities (containing species such as *Stratiotes aloides, Potamogeton natans, Potamogeton prealongus, Potamogeton lucens, Utricularia vulgaris* and a variety of Charophytes) to an extensive area of open 'swamp' dominated by *Schoenoplectus lacustris, Typha angustifolia* and *Nymphaea alba* and thence to a closed reed swamp of *Phragmites australis* (Pallis, 1911) (Fig. 8.2).

During the first half of the 20th century the open water flora began to be replaced by faster growing, taller water weeds, such as *Myriophyllum spicatum, Potamogeton pectinatus* and *Cerato-phyllum demersum* (George, 1970). These became so prolific that there was widespread concern locally that the open water would become so choked that the movement of boats would become impossible and the newly-growing holiday industry ruined. By the end of the 1950s however the submerged aquatic plants, together with much of the fringing swamp, had completely disappeared from all but a handful of lakes, to be replaced by substantial phytoplankton populations and an abrupt transition to the exposed edge of the original peat basin.

The open water plants and surrounding 'open swamp' provided cover, egg-laying sites and a source of food for a wide variety of animals, particularly invertebrates (Mason and Bryant, 1975) and the loss of these communities has caused a general reduction in diversity of wildlife in the area (Mason, 1976). Concern over the ecological quality of this important wetland initiated a systematic study of the ecology of the area the initial conclusions of which were summarised by Anon. (1982b). This work has provided a detailed understanding of the effect of eutrophication on shallow lakes and the scientific foundation for a programme of restoration which began in 1977 and is still continuing in 1990.

8.3 EUTROPHICATION OF THE BROADS

It is now clear that the transition from shallow weedy lakes, rich in associated animals to a phytoplankton-dominated lake of low diversity was associated with an increase in eutrophication. Three phases of the open water community of these lakes have been distinguished (Moss *et al.*, 1979b). The initial phase, 'Phase 1', dominated by charophytes and *Najas* in clear water, is now restricted to very few lakes and is associated with annual mean total phosphorus concentrations of less than 50–60 µgP/l (Table 8.1). The second phase, 'Phase 2', of taller-growing 'rank' species, such as *Myriophyllum spicatum* and *Potamogeton pectinatus*, occurs in lakes with slightly higher phosphorus concentrations, up to 100–125 µgP/l. The third phase, 'Phase 3', is found at greater phosphorus concentrations than this. No submerged aquatic plants are found and the lakes are dominated by phytoplankton (Fig. 8.3). Examination of dated sediment cores provides further evidence for a link between eutrophication and these changes. Deeper sediments laid down early in the lakes' history have much lower phosphorus concentrations than recent sediments. In Barton Broad pre-1800 sediment has a phosphorus content of approximately 0.5 mg P/g dry wt, this increases to 1.0 mg P/g in the 19th century and in the

Table 8.1 Annual mean lake total phosphorus and chlorophyll 'a' concentration (μg /l) in 1988 (National Rivers Authority, unpublished data)

Lake	Total P	Chl 'a'	Dominant plant
Martham South Broad	25.0	9.7 Phase 1	Charophyte
Upton Broad	54.0	8.2 Phase 1	*N. marina*
Horsey Mere	59.0	43.0 Phase 2	*M. spicatum*
Hickling Broad	106.0	66.0 Phase 2	*P. pectinatus*
Wroxham Broad	82.0	65.8 Phase 3	Plankton
Barton Broad	129.0	149.2 Phase 3	Plankton
Ranworth Broad	127.0	156.1 Phase 3	Plankton
Barton Broad	309.0	161.5 Phase 3	Plankton
South Walsham Broad	346.0	168.0 Phase 3	Plankton

most recent deposits ranges from 2 to 7.5 mg P/g (Moss, 1980). Changes are also found in the number and species of diatom frustules found in the sediment. In the early deeper sediments epiphytic diatom taxa predominated (predominantly *Fragilaria* but including *Epithemia*, *Cocconeis* and *Achnanthes* spp.). These taxa increased in abundance until c.1924 but after this time planktonic taxa (centrales, mainly *Cyclotella* and *Melosira*) became abundant (Moss, 1977). Sediment cores from other broads have shown similar changes (Moss *et al.*, 1979a) and in addition remains of molluscs and plant-associated zooplankton are more abundant in deeper sediment, while small planktonic zooplankton typical of eutrophic lakes such as *Bosmina longirostris* occur in the more recent deposits (Stansfield *et al.*, 1989).

8.4 MECHANISMS OF CHANGE

The increasing fertility of the Broadland lakes has almost certainly been associated with a general increase in population and agricultural activity. Osborne and Moss (1977) demonstrated that the opening of a sewage treatment works in 1924 coincided with a marked increase in phosphorus sedimentation in Barton Broad. As the water became more fertile, primary productivity will have increased. Initially this was expressed in the growth of aquatic plants as the lake changed from Phase 1 to Phase 2 but was subsequently overtaken by the growth of phytoplankton in the final Phase 3 state. Although phytoplankton can cause considerable shading, light absorption by phytoplankton is frequently insufficient to prevent net rooted plant photosynthesis at the sediment surface and by mid-summer many submerged plant species have produced a substantial proportion of their photosynthetic biomass in the upper 20 cm of the water column (Fig. 8.4) (Phillips and Moss, 1978). Under these conditions it seems unlikely that competition for light would cause the loss of plant populations in such shallow lakes and the mechanism for their demise is probably more complex. Eminson and Phillips (1978) showed a reduction in growth of *Najas marina* with epiphyte growth at high nutrient concentrations in the laboratory, and Phillips *et al.* (1978) postulated that at low nutrient concentrations rooted aquatic plants, which could obtain nutrients from the sediment, were able to prevent epiphytic growth by organic secretions. At higher nutrient levels when both phytoplankton and epiphytic growth was greater the combined effect of shading by these algae could reduce the plants' capability of producing algicidal secretions, reduce growth rate and eventually lead to the loss of the submerged plants.

The reasons for the loss of fringing swamp communities is less well understood although this can also be traced to the effects of eutrophication. Fringing reed occurs in one of two forms, a rooted mat, where the lake basin forms a shallow slope and a floating 'hover', which grows out from the rooted material at the edge of the lake basin where the side is steep (Crook *et al.*, 1983). Crook and her colleagues were able to demonstrate that the majority of loss of reed had occurred from areas of hover, and not from the directly rooted material, and that the extent of reed loss was positively correlated with the maximum nitrate concentrations of the water. They also demonstrated a positive effect of nitrate enrichment on above-ground growth and a negative affect on below-ground growth and postulated that this would lead to instability of hover reed, thus increasing its susceptibility to wave action. Kloetzi and Zust (1973) suggested that increased nitrogen availability leads to a decrease in mechanical stability of reed stems through decreased production of schlerenchyma fibres in the stems. There is thus evidence to suggest that the loss of some communities of fringing emergent vegetation is due to eutrophication.

8.5 EFFECTS OF CHANGE

These changes have had considerable consequences for the ecology of the open water communities of these shallow lakes, the majority of the effects resulting indirectly from the loss of aquatic plants.

Bird populations have changed, particularly those that depend on aquatic plants for food. Hickling Broad retained its submerged aquatic flora until the early 1970s (Phillips *et al.*, 1978) and the loss of these plants coincided with a substantial reduction in the lakes' mute swan (*Cygnus olor*) population. Three hundred and twenty non-breeders in 1961 decreased to 69 in 1978 (Taylor, 1978). Similar data from the British Trust for Ornithology show a decrease in tufted duck (*Aythya fuligula*) and pochard (*Aythya ferina*) on Hickling Broad, from 160 and 66 respectively in 1970, to 31 and 46 between 1975 and 1979 (George, in press).

Invertebrate diversity of the open water benthos is very low and most of the lakes are dominated by oligochaete worms of the family tubificidae: *Limnodrilus hoffmeisteri* and *Potamothrix hammoniensis* together with chironomid larvae of the genus *Chironomus* (Table 8.2). Molluscs are particularly poorly represented, the open water benthos containing the swan mussel (*Anodonta cygnea*) and in a few sites where the lake bottom contains less silt *Bythinia tentaculata* and *Valvata piscinalis* are found. In contrast lakes that still retain submerged plant populations are more diverse and contain several mollusc species (Mason and Bryant, 1975). Wortley (1974) demonstrated that this was a direct result of lack of habitat diversity and showed than when artificial plants constructed from polypropylene were introduced into the lakes large number of invertebrates established populations. Although enrichment implies an increase in primary productivity which should result in a greater potential food supply for benthic detritivores, secondary production has not increased (Mason, 1977) and diversity has decreased dramatically. The littoral fringe of many of the broads contain a much wider diversity of invertebrates and it is clear that the very fluid sediment characteristic of the open water of these lakes does not provide a suitable habitat for many of these animals.

This lack of diversity is also reflected in the fish populations. The Broads have traditionally been important for angling and in the past for substantial catches of roach, rudd, perch and pike (Moss *et al.*, 1979b). Estimates of fish standing crop from the Broads are very variable (Table 8.3) but generally reveal a low biomass (approximately 100 kg/ha) and a truncated age distribution dominated

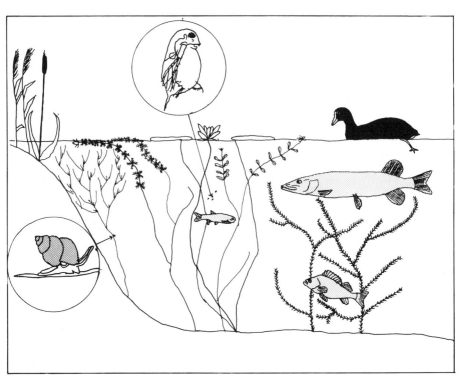

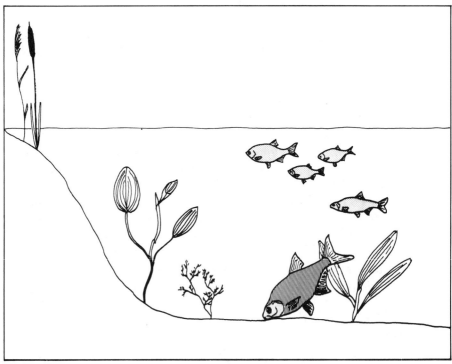

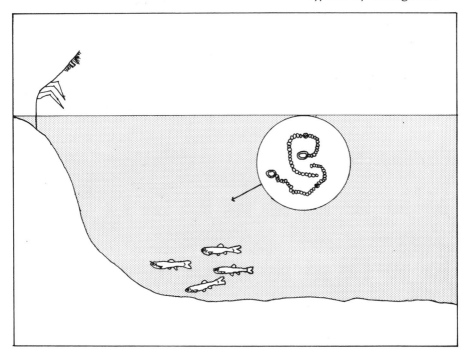

Figure 8.3 Diagrammatic representation of the three phases of entrophication in the Broads. See text for explanation.

numerically by roach less than 4 years old.

Growth rates of the young fish are relatively high but it seems likely that there is an inadequate food supply for older fish. Young roach are able to exploit the high primary production of phytoplankton *via* an abundant zooplankton but older roach and perch appear to depend on aquatic weeds for cover, avoiding predation and habitat for prey. The inbalance in age distribution is exacerbated by lack of predation on young fish. In clear water-weed dominated lakes, pike are likely to limit fish densities but in the turbid weed-free water typical of the Broads, pike populations provide little predation pressure (Grimm, 1989). Unlike similar lakes in the Netherlands the pike-perch (*Stizostedion lucioperca*) does not occur in the Broads and this may explain the numeric dominance of these lakes by young roach rather than bream which occur in the Netherlands. Bream populations, although not numerically important, have a wider age structure and often contribute significantly to fish biomass. This is probably due to their

Table 8.2 Macro-invertebrate taxa found in the open water and the littoral fringe of three broads, summer 1988. (Anglian Water unpublished data)

Taxon	Alderfen Broad	South Walsham Broad	Hoveton Gt. Broad
Open Water			
Potamothrix hannoniensis	va	c	a
P. bavaricus			c
Limnodrilus hoffmeisteri		c	
Ostracods	c		
Chironomidae	a	c	c
Ceratopogonidae			p
Chaoborus			p
Hydracarina	p		
Littoral			
Polycelis tenuis	p		
Dugesia tigrina			c
Tubificidae	p		p
Erpobdella octoculata	p	p	
Helobdella stagnalis	p	p	p
Theromyzon tessulatum	p		
Asellus aquaticus	c	p	
Asellus meridianus	c	p	p
Crangonyx pseudogracilis	c		c
Gammarus zaddachi		a	
Neomysis integer		a	
Palaemonetes varians		p	
Hydracarina		c	c
Cloeon dipterum	p		p
Limnephilidae		p	
Cyrnus flavidus	p		
Phryganea obsoleta	p		
Sialis lutaria	p		p
Elminthidae	p		
Dytiscidae		p	
Notonecta glauca	p	p	p
Corixidae	c	a	a
Ischnura elegans	c		
Erythroma nais	p		
Coenagrion		p	
Ceratopogonidae			p

Table 8.2 (contd.)

Taxon	Alderfen Broad	South Walsham Broad	Hoveton Gt. Broad
Chironomidae	a	p	c
Tanypodinae	c		c
Phytoptera	p		
Valvata piscinalis	p	p	p
Bithynia tentaculata	c	p	p
Physa fontinalis	p	p	
Limnea peregra	p	p	p
Planorbis carinatus	p		
P. vortex	p		
P. albus			p

va = very abundant, a = abundant, c = common, p = present

Table 8.3 Mean biomass of fish from a variety of Broads (kg/ha) (National Rivers Authority)

Broad	Year	Mean biomass (kg/ha)
Barton Broad	1979	77.8
Barton Broad	1985	177.8
Barton Broad	1988	319.3
Ranworth Broad	1984	44.5
Ranworth Broad	1987	54.2
Hickling Broad	1986	123.0

ability to feed successfully on chironomids in very fluid sediments (Lammens, 1989) and thus exploit the secondary production of these lakes.

In some of the lakes (Hickling Broad, Horsey Mere) their proximity to the North Sea results in slightly brackish conditions and at these sites the alga *Prymnesium parvum* frequently develops. This alga produces a fish toxin and has been responsible for the death of large numbers of fish and probably other gill-bearing invertebrates (Wortley and Phillips, 1987). Elsewhere phytoplankton is typical of eutrophic lakes with diatoms (both centric forms such as *Stephanodiscus hantzschii*, *Cyclotella meneghiniana* and pennates such as *Synedra* spp., *Nitzschia acicularis* and *Diatoma elongatum* dominating during spring and autumn with small green algae belonging to the Chlorococcales and cyanobacteria species (*Oscillatoria limnetica*, *Anabaena* spp. and *Aphanizomenon flos-aquae*) during the summer). The relative importance of algal groups is strongly related to the hydrological regime. In lakes that receive little riverine water exchange, substantial populations of cyano-bacteria develop in early summer, remaining until flushed out by increased river discharge in the autumn. In well-flushed lakes the phytoplankton tends to have a greater proportion of faster growing small chlorococcales (green algae) or bacillariophyceae (diatoms).

8.6 RESTORATION OF BROADLAND

8.6.1 Sources of nutrients

The obvious way to reverse eutrophication is to reduce the nutrient supply. It is now widely accepted that phosphorus rather than nitrogen is the key element to control in a programme of eutrophication abatement. This is partly a pragmatic choice, in that nitrogen comes from water draining agricultural land and would be very difficult to control, whereas phosphorus originates from point sources and can easily be removed prior to discharge. In Broadland phosphorous has been shown to be the primary limiting nutrient (Osborne, 1978; Watson, 1981), although during the summer, when phytoplankton crops are very high, nitrogen is often undetectable in the lake water resulting in temporary limitation of algal growth. Nitrogen limitation may mean that reducing the supply of phosphorus will initially have only a limited effect in reducing phytoplankton standing crop, until a new threshold of phosphorus limitation is achieved.

8.6.2 Phosphorus removal from sewage treatment works

The principal source of phosphorus to the majority of Broadland lakes is from sewage treatment works effluent. Osborne (1981) in a study of the River Ant and Barton Broad, showed that approximately 90% of phosphorus discharged to the catchment of the River Ant was from sewage treatment works, with only 10% from other catchment sources. In the larger catchment of the River Bure 78% of phosphorus was derived from sewage sources (Moss *et al.*, 1988).

In both these studies the amount of phosphorus discharged into the river from sewage treatment works was substantially more than that passing a point in the river several kilometres downstream, suggesting that there was a considerable loss of phosphorus to the biota or the river sediment. In the River Ant this amounted to 20% of the sewage contribution. This loss of phosphorus may be due to direct sedimentation of particulate material or uptake by chemical or biological mechanisms. Whatever the mechanism it poses considerable problems when making management recommendations regarding the most important sources of phosphorus, as simple studies do not provide a way of apportioning the loss of phosphorus between different sewage effluent discharges, or to the contributions from the catchment.

Phosphorus removal is carried out by adding ferric sulphate to the effluent stream just prior to final settlement. The process is capable of removing at least 90% of soluble phosphorus provided dose rate is carefully controlled to match the amount of phosphorus in the effluent. Initially this was achieved by dosing ferric sulphate at a fixed rate during periods of average to high flow, but stopping dosing overnight when flow was low. However, the dosing rate is now controlled using a redox control system. The addition of ferric sulphate increases the redox potential of the effluent and by measuring the redox potential of the dosed effluent stream the amount of chemical addition can be very precisely regulated (Kerrison *et al.*, 1989).

8.6.3 The reduction of phosphorus input to the River Ant and Barton Broad

The control of eutrophication in Broadland began in 1977 with the reduction in phosphorus input to the River Ant upstream of Barton Broad, the second largest of the Norfolk Broads. In 1986 this was extended to include major sewage treatment works in the River Bure catchment. The target total phosphorus concentration for the inflowing water was set at 100 µg/l, a value low enough to allow Phase 2 plant communities to exist (Table 8.1). Lower values, where Phase 1 conditions could be created, could not be achieved due to the input of phosphorus from diffuse sources within the catchment.

The catchment of the River Ant contains 16 sewage-treatment works; most of these are very small works, and phosphorus removal was undertaken only at the larger works (Phillips, 1984).

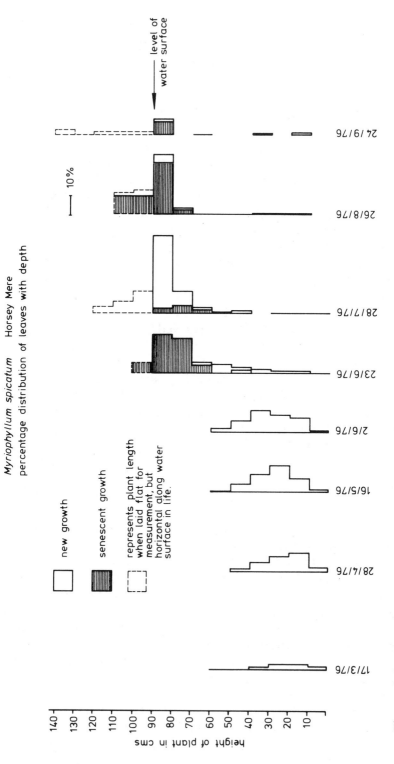

Figure 8.4 Growth and biomass distribution of *Myriophyllum spicatum* in Horsey Mere.

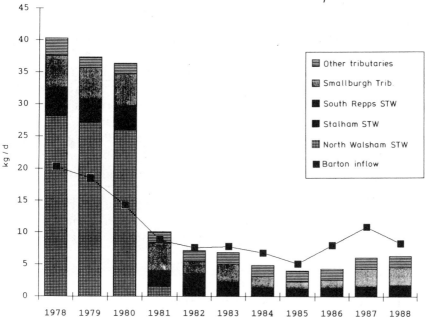

Figure 8.5 Reduction of phosphorus inputs to the inflow stream of Barton Broad after removal of phosphorus from the major sewage works effluent.

The effect of this was to reduce the amount of phosphorus discharged to the River Ant upstream of Barton Broad by at least 90%. Monitoring of the river at the mouth of Barton Broad however showed that phosphorus loads were only reduced by about 50% (Fig. 8.5) and mean phosphorus concentrations in Barton Broad were very variable and often only slightly lower than pre-phosphorus removal (Fig. 8.6). These results demonstrate the difficulties of predicting the effectiveness of reducing phosphorus loads from particular phosphorus sources.

8.6.4 The role of the sediment in the nutrient budget of the River Ant

It is clear that a considerable quantity of phosphorus enters the sediment in the river and that this sediment phosphorus load is carried downstream and eventually enters Barton Broad (Osborne and Moss, 1977; Osborne, 1981). How far this sediment moves, and under what conditions is not understood but in the navigable sections of the river it is clear that resuspension of sediment by motor boats is an important factor.

The River Ant upstream of Barton Broad is intensively used by motor boats. The river is narrow and as boat movement builds up sediment is resuspended from the river bed and maintained in suspension until the evening when boat activity is less (Hilton and Phillips, 1982). This sediment contains phosphorus and clearly adds to the measured total phosphorus flux moving down the river. Water samples collected before boats begin to move (at about 6.00 am) contain substantially less particulate phosphorus than samples collected later in the day. This particulate phosphorus load partly accounts for the higher phosphorus loads in the river compared to that discharged from sewage treatment works, but local deposition of phosphorus downstream of effluents in the non-navigable section of the river probably reduces their effective relative contribution in comparison with diffuse input from the catchment.

8.7 LAKE SEDIMENT AS A SOURCE OF PHOSPHORUS

During the summer dissolved phosphate is usually undetectable in Barton Broad due to rapid uptake by phytoplankton; however, a marked peak of dissolved phosphate is often observed during June (Fig. 8.7). This is rapidly taken up by phytoplankton and the total phosphorus concentration of the lake water is often greater than that of the inflowing river. Measurements of total phosphorus input and output have demonstrated that during periods in the summer there is a net release of phosphorus to the lake water from the sediment. This has been widely noted in other lakes (Marsden, 1989) and is now recognized as a major reason why many shallow lakes fail to respond quickly to phosphorus control measures (Sas, 1989).

The magnitude of phosphorus input from lake sediments can be estimated from input/output budgets; however, for Barton Broad the interpretation of these budgets is complicated by the nature of the particulate phosphorus which forms a substantial proportion of both the inflowing and outflowing phosphorus. The inflowing river water contains particulate phosphorus primarily derived from resuspended sediment material, while particulate phosphorus leaving the lake is mostly composed of phytoplankton. Thus although the phosphorus budgets can demonstrate a net loss or gain of phosphorus from the lake they cannot determine the amount of deposition of riverine sediment-bound phosphorus or the release of

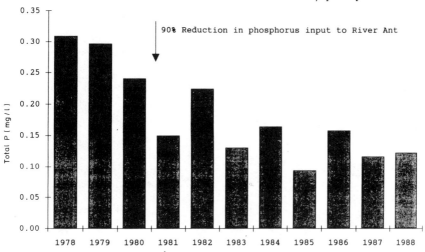

Figure 8.6 Annual mean total phosphorus in Barton Broad before and after phosphorus removal from sewage effluents in the river Ant.

soluble phosphorus in a form in which it can be incorporated into the phytoplankton.

Phosphorus release rates from the sediment of Barton Broad are probably very high. The sediments of Barton Broad are relatively rich in total phosphorus (Table 8.4) and can contain relatively high phosphate concentrations in the pore water. Experimental measurements of phosphorus release from undisturbed sediment cores in the laboratory, using a flow-through system, incubated at environmental temperatures, have demonstrated that during the winter the sediment of Barton Broad absorbs phosphorus, but from May until August phosphorus release occurs (Fig. 8.8). Peak release rates can be as high as $130 \, mg/m^2/day$ and although these experiments probably represent maximal rates of release, they demonstrate that loss of phosphorus from the sediment of Barton Broad can be very high. A number of laboratory incubations carried out on sediment from Barton Broad during 1988 gave an annual average release rate of $8.0 \, g/m^2/annn$, which is similar to the estimated external load to the lake prior to the control of phosphorus (Phillips & Jackson, 1989).

These very high rates of release cannot be accounted for by diffusion kinetics using measured phosphorus gradients in the interstitial pore water and it is clear from experimental work that benthic chironomid larvae considerably enhance the rate of release

Table 8.4 Total, inorganic and organic phosphorus content of Barton Broad sediment (30/11/88)

Sample depth (cm)	Total inorganic P (mg/g)	Organic P (mg/g)	Total P (mg/g)
1	0.91	0.43	1.34
2	0.80	0.53	1.33
3	0.84	0.63	1.47
4	0.69	0.67	1.36
5	0.59	0.55	1.14
6	0.55	0.40	0.95
7	0.46	0.04	0.50
8	0.49	0.14	0.63
9	0.40	0.43	0.83
10	0.37	0.34	0.71

(Phillips and Jackson, in press). The mechanism by which this happens is not known. Andersen and Jensen (in press) have shown similar results and suggest that the chironomids enhance the decomposition of the surface sediments, but simple bulk movement of pore water containing relatively high phosphorus concentrations may be enough.

8.8 CHANGES IN THE STORAGE OF PHOSPHORUS IN LAKE SEDIMENT FOLLOWING PHOSPHORUS CONTROL

The net annual amount of phosphorus stored in the sediment of a lake (sed.P g/m²/ann) might be expected to be reduced following the reduction of external load (Sas, 1989). Data from Barton Broad for 1978–88 apparently confirms this (Fig. 8.9) with a positive value prior to the introduction of full phosphorus removal in 1980, followed by an unstable period in which the lake appears to be in a transition state. However, if data collected by Osborne (1981) are included for the years 1975–76, it is apparent that low or negative values for sed.P can occur at high phosphorus loads. There was an exceptionally dry summer in 1976; substantial peaks of dissolved phosphate were observed in the lake and release of phosphorus from the sediment clearly occurred resulting in a large algal biomass. The majority of this phosphorus was retained by

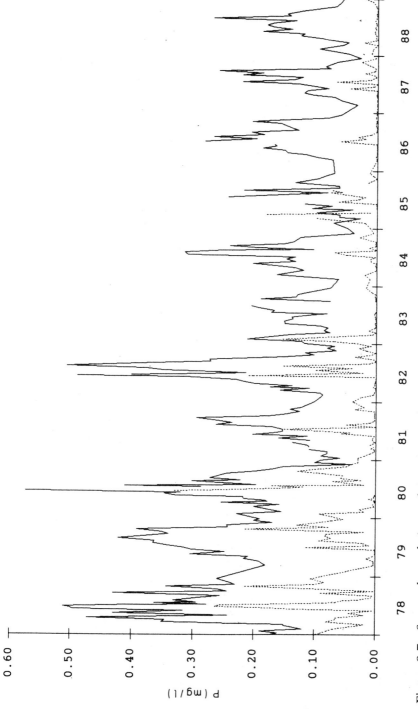

Figure 8.7 Seasonal trends in total (continuous line) and soluble (dotted line) phosphorus concentrations in Barton Broad.

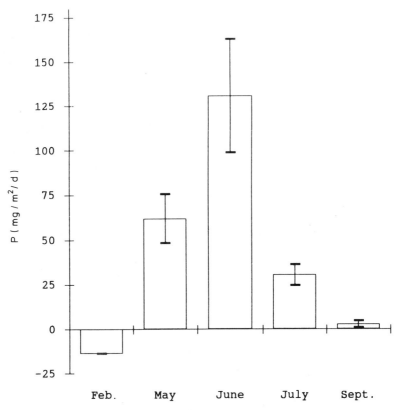

Figure 8.8   Seasonal pattern of phosphorus release from sediment cores taken from Barton Broad.

phytoplankton and lost from the lake in the autumn when river flow increased and washed phytoplankton from the lake. Thus when summer flow is low, a larger amount of phosphorus from the lake sediment can be lost from the lake via the phytoplankton.

A plot of phosphorus sedimentation (sed.P) against summer river discharge (April–September mean) demonstrates the importance of river flow in determining the retention or loss of phosphorus from the lake (Fig. 8.10). The effect of phosphorus control post-1980 is clearly seen as a decrease in net sedimentation of about 4 g/m^2/ann (8 kg/lake/day) which is similar to the measured decrease in load to the lake. These results suggest that the rate of phosphorus loss from the sediment of Barton Broad has remained very similar over the last ten years despite the introduction of phosphorus control measures.

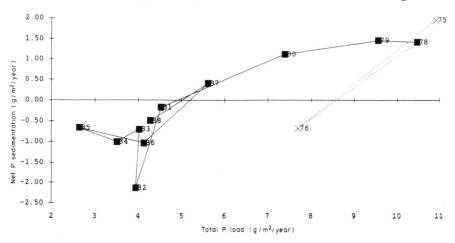

Figure 8.9 Annual mean of daily phosphorus sedimentation rates in Barton Broad.

In addition to the release of soluble phosphorus from the sediment it is possible that phosphorus in resuspended sediment may also become available. Approximately 40% of phosphorus in the upper 5 cm of Broads sediment is extractable using a bicarbonate–dithionate solution, and this is probably bound to ferric iron. At high pH this may become soluble due to ligand exchange mechanisms (Stauffer and Armstrong, 1986). Thus in mid-summer when the pH of some of the Broads can be in excess of 9.0 as a result of high phytoplankton crops, soluble phosphorus may become available for phytoplankton growth.

8.9 BIOLOGICAL RESPONSE TO REDUCED NUTRIENT LOADING

Due to the large reservoir of phosphorus contained in the sediment of both the River Ant and Barton Broad, phosphorus control in the catchment of Barton Broad has had only limited success in reducing lake phosphorus concentrations and consequently the changes in the lake biota have been minimal.

The average phytoplankton standing crop has been reduced and there is some evidence that algal species composition has changed slightly, the main effect being a reduction in the number of centric diatoms and slight decreases in numbers of cyanobacteria, particu-

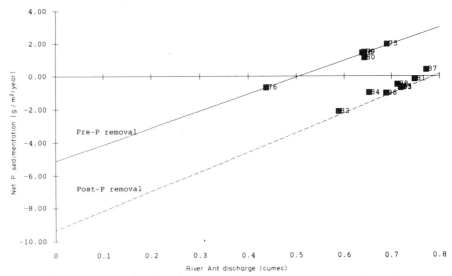

Figure 8.10 Annual phosphorus sedimentation compared with river discharge (hence flushing rate) for years before and after phosphorus removal from the inflow.

larly heterocystous species such as *Aphanizomenon flos-aquae*. However, algal standing crop is still substantial and there is no indication that phosphorus levels have been reduced sufficiently to allow the re-establishment of submerged aquatic plants despite extensive planting trials using a variety of species. There is a clear relationship between mean annual chlorophyll a concentration and total phosphorus (Fig. 8.11) and if the log transformed data are fitted to a linear regression the exponent B in the relationship chlorophyll 'a':total phosphorus takes the value 0.74, which is similar to that quoted by Sas (1989) (0.77) in a study of eighteen European lakes. This suggests that the Broadland lakes behave in a very similar way to other temperate lakes. To reduce chlorophyll 'a' concentrations to values where the water is sufficiently clear to allow plant growth, probably less than 50 µg/l P, the average phosphorus concentrations would need to be reduced to about 30 µg/l . This would be difficult to achieve without further measures to control diffuse sources, which at present ensure that background phosphorus concentrations in the inflow to the lake are greater than 50 µg/l. Moreover, phosphorus release from the sediment is likely to continue to provide a substantial source within the lake unless action such as sediment removal is taken.

8.10 NUTRIENT REDUCTION BY ISOLATION

A second approach to the reduction of nutrient input to these lakes has been the diversion of inflowing water and therefore its nutrients. This can be carried out only where navigation does not occur and the inflow stream is sufficiently small to allow the engineering work to be feasible. The technique has the advantage of not requiring expensive treatment of effluents and has been used in a number of small lakes in Broadland. In one of these lakes, Alderfen Broad, isolation initially appeared to be very successful. After two years phytoplankton was greatly reduced and a submerged plant population had been re-established in the Broad. However, in the following three years these plants disappeared and phytoplankton populations again dominated the lake (Moss *et al.*, 1986). During the initial phase of plant re-establishment, isolation of the lake severely restricted input of both nitrogen and phosphorus and appears to have limited the release of phosphorus from the sediment. The cause of this reduction in release of sediment-derived phosphorus is not known but it may have been the result of decreased organic matter input from spring phyto-plankton populations. However, once plants were re-established phosphorus release from the sediment began to increase again, perhaps due to the input of organic material from the plants, with the eventual effect of phytoplankton again dominating the Broad (Moss *et al.*, 1986). The results from Alderfen demonstrate the limitation of isolation as a technique; a lake with little or no water exchange is likely to accumulate nutrients and via nutrient re-mobilisation to return inevitably to phytoplankton dominance.

In both the restoration examples given, sediment derived phosphorus had been highlighted as the primary reason for failure of the lakes to recover. The work at Alderfen Broad suggested that at least in the short term a reduction of nutrient input resulted in less release of phosphorus from the sediment. In the case of Barton Broad however, available evidence suggests that phosphorus release has not decreased. This may be due to the fact that the initial reduction in nutrient supply at Barton Broad was insufficient to lower organic input to the sediment, and hence influence phos-phorus release, or that release rates are not controlled in this way and that a combination of other circumstances lead to the changes in Alderfen Broad. These issues have still to be resolved and an understanding of the mechanisms controlling release of phosphorus from the sediment of the broads are vital if restoration of these lakes is to be achieved by nutrient control.

8.11 SEDIMENT REMOVAL AND LAKE ISOLATION

One way to overcome the release of phosphorus from the sediment is by removing it from the lake. This is an expensive process but in situations where water has become so shallow that the lake can no longer be used, the process can be justified. Two lakes in the River Bure valley have been treated in this way by the Broads Authority (Cockshoot Broad in 1982 and Belaugh Broad in 1987). The removal of sediment from Hoveton Little Broad was started in 1990 and will be completed during 1991. The cost of the removal of sediment from Cockshoot Broad was £50,000 although the economies of scale mean that the costs for the Hoveton Little Broad are estimated as £23,000, about half the unit cost. Cockshoot Broad was also isolated from the River Bure, thus preventing the input of nutrient-rich water, but the introduction of phosphorus control in the catchment of the River Bure in 1986 has allowed Belaugh Broad to remain connected to the main river. At both sites phytoplankton growth has been moderate and although a diverse submerged plant population was quickly established in part of Cockshoot Broad (Moss *et al.*, 1986) the majority of the open water of these two lakes have not been colonised by plants. At both sites phosphorus concentrations have remained below 100 µg/l for the majority of the year and experimental cores have confirmed that phosphorus release does not occur. The failure of plant colonisation is not understood but may be due to a combination of lack of a sufficient inoculumn of plant material and to excessive grazing by water birds such as coot (*Fulica atra*), or simply to the fact that phytoplankton populations are still preventing adequate light transmission.

8.12 BIOMANIPULATION AS A RESTORATION TECHNIQUE

It is clear that phosphorus control within the catchment may reduce phosphorus levels to average values ranging from 60 to 120 µg/l, although phosphorus from the sediment may delay or even prevent these levels from being achieved. A question that remains to be answered is whether aquatic plant populations can be re-established at these phosphorus concentrations. Balls *et al.* (1989) have suggested that the phytoplankton-dominated state and the aquatic plant-dominated state can be alternatives over this range of phosphorus concentrations. A critical factor allowing this is the

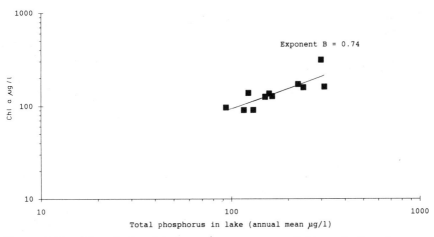

Figure 8.11 The relationship between mean annual chlorophyll and mean annual total phosphorus in Barton Broad.

effect of grazing by large-bodied zooplankton on the phytoplankton.

The ability of herbivorous zooplankton to remove phytoplankton from the water is controlled by their filtration rate, which in turn is usually proportional to body size raised to a power of 1.5 to 3 (Peters, 1984). The majority of the Broads contain a zooplankton community that is dominated for most of the summer by small-bodied cladocerans, particularly *Bosmina longirostris*, copepods and rotifers. Grazing by these small zooplankton provides little potential for the removal of significant amounts of phytoplankton. This contrasts with the situation in the late spring when relatively high population densities of larger cladocerans (e.g. *Daphnia hyalina*) occur (Phillips, 1984). These population peaks are often coincident with declines in phytoplankton and even when nutrient concentrations remain high, relatively clear water can occur for a few weeks during the period of high cladoceran densities.

The decline in the populations of these zooplankton is almost certainly caused by predation by young fish. Studies at Alderfen Broad (Cryer *et al.*, 1986; Townsend and Perrow, 1989) have demonstrated that young-of-the-year roach can rapidly deplete cladoceran prey. As alternative prey is very scarce, due to the absence of aquatic plants and their associated fauna, the loss of the prey tends to reduce fish fecundity and in the following year predation pressure by young fish is less. This results in a two-year cycle of zooplankton and roach abundance. In Alderfen Broad

Figure 8.12 The results of enclosure experiments protecting zooplankton from fish predation in a Norfolk Broad devoid of macrophytes, 1990.

convincing evidence of an effect on phytoplankton populations has not been obtained, perhaps due to other major perturbations brought about by isolation.

Elsewhere, evidence is more convincing, however. Small cages (1 metre × 2 metres) made of a fine nylon mesh (aperture 100 μm) were introduced into Barton Broad and planted with a variety of submerged aquatic plants. Similar plantings were carried out in an area outside the cages and phytoplankton and zooplankton populations monitored. The cages contained no fish and inside very large cladoceran populations were established. Phytoplankton densities were reduced, although water movement through the mesh ensured that chlorophyll levels remained moderately high. A diverse aquatic plant population developed inside the cages although no plants survived outside (Fig. 8.12).

The role of fish/zooplankton interactions was also shown in a series of experimental ponds (Balls *et al.*, 1989; Irvine *et al.*, 1989). In these ponds aquatic plant populations were shown to be extremely resilient to artificial phosphorus and nitrogen fertilisation (60 g N/m^2; 0–2.6 g P/m^2) probably as a result of large *Simocephalus* populations maintaining low phytoplankton. Similarly in ponds artificially cleared of plants, phytoplankton became abundant only when small roach were introduced to the ponds and *Daphnia* populations were reduced.

The effectiveness of cladocerans at controlling phytoplankton was also demonstrated in Hoveton Great Broad (Timms and Moss, 1984). The main basin of this lake is weed-free and dominated by phytoplankton although a subsidiary basin, Hudsons Bay, has a rich stand of water lilies (mostly *Nuphar lutea*) and very low phytoplankton populations (<10 μg/l) throughout most of the early summer. Both basins were fed by nutrient-rich water from the River Bure (prior to phosphorus control in this catchment). Phosphorus concentrations were similar in both basins and often much greater than 100 μg/l. The difference in phytoplankton was shown to be a result of much greater zooplankton grazing by a combination of weed-associated cladocerans and larger-bodied planktonic species. Timms and Moss were able to demonstrate that co-existence of the larger-bodied cladocerans with a fish population was due to the provision of day-time refuges for the cladocera among the lilies and efficient night-time feeding in the open water.

The association of grazers with weed beds, particularly around the edge of these lakes, may be a very important factor in maintaining clear water conditions in the open water of the lake. Photographs of the Broads taken in the last century reveal an extensive area of emergent plants extending into the open water of these lakes and this may have provided a refuge for a wide variety of zooplankton and a consequent grazing pressure on phytoplankton growth (Fig. 8.2). These weed beds would also have provided ideal habitat for pike and with a more diverse invertebrate fauna it might be expected that fish biomass would be spread over a much wider age range with consequently less predation pressure on the grazing cladocera.

8.13 THE RESTORATION OF BROADLAND IN THE CONTEXT OF TWO STABLE COMMUNITIES

It is no longer possible to see the restoration of Broadland as a simple transition from nutrient rich phytoplankton-dominated lakes to diverse weed-filled lakes simply by the control of phosphorus discharged to the waterways. The broadland lakes of the 19th century certainly had a lower nutrient status, probably much lower than can now be achieved without radical change to surrounding land use, and their dominance by aquatic plants was probably an inevitable consequence of factors such as greater access to sediment nutrients, protection of grazing zooplankton and perhaps allelopathy. Conversely, at very high nutrient loads phytoplankton may be the inevitable result; zooplankton grazing pressure in the open water is minimal, and where filamentous cyanobacteria dominate, ingestion by the zooplankton may be impossible or ineffective. However, in between these two extremes either community may exist. When nutrient supply is increasing, stabilising mechanisms enable the plant community to survive, while during the recovery from enrichment plankton-stabilising mechanisms prevent the re-establishment of these plants.

Thus although phosphorus control is still a prerequisite to lake recovery it seems unlikely that phosphorus loads can be reduced sufficiently to enable the inevitable establishment of aquatic plants without additional measures, particularly when the effect of sediment-derived phosphorus is taken into account. If grazing zooplankton are an important component in the stabilising mechanisms of weed dominated lakes, as available evidence suggests, then removing fish populations could provide the trigger needed for complete lake recovery. Lake recovery has been achieved in this way in the Netherlands (Van Donk *et al.*, 1989; Meijer *et al.*, 1989) and large-scale experiments in the Norfolk Broads are planned. The permanent and complete removal of fish would however be as impractical in the Broads as it would be inappropriate, due to the interconnections of the lakes with navigable river and the importance of fishing as a recreational activity. However temporary removal of fish from the littoral fringes of some of these lakes may provide a means of re-establishing the aquatic plant communities around the fringe of the lake. Once established, and given a reduced phosphorus loading, they may provide the habitat diversity that will lead to a change in the community structure of these shallow lakes and their complete restoration.

References

Adams, M.S. and Prentki, R.T. (1982) Biology, metabolism and functioning of littoral submersed weedbeds of Lake Wingra, Wisconsin, USA: a
summary and review. *Archiv für Hydrobiologie Supplement*, **62**,
333–409.

Addiscott, T. (1988) Farmers, fertilisers and the nitrate flood. *New
Scientist*, **120**, 50–4.

Ahl, T. (1975) Effects of man-induced and natural loadings of phosphorus
and nitrogen on the large Swedish lakes. *Verhandlungen Internationale
Vereinigung für Theoretische und Angewandte Limnologie*, **19**, 1125–32.

Ahl, T. (1988) Background yield of phosphorus from drainage area and
atmosphere. *Hydrobiologia*, **170**, 35–44.

Ahlgren, I. (1972) Changes in Lake Norrviken after sewage diversion.
*Verhandlungen Internationale Vereinigung für Theoretische und Ange-
wandte Limnologie*, **19**, 355–61.

Ahlgren, I. (1978) Response of Lake Norrviken to reduced nutrient
loading. *Verhandlungen Internationale Vereinigung für Theoretische
und Angewandte Limnologie*, **20**, 846–50.

Ahlgren, I. (1980) A dilution model applied to a system of shallow
eutrophic lakes after diversion of sewage effluents. *Archiv für Hydrobiologie*, **89**, 17–32.

Ahlgren, G. (1988) Phosphorus as a growth-regulating factor relative to
other environmental factors in cultured algae. *Hydrobiologia*, **170**,
191–210.

Ahlgren, I., Frisk, T. and Kamp-Nielsen, L. (1988) Empirical and
theoretical models of phosphorus loading, retention and concentration
vs. lake trophic state. *Hydrobiologia*, **170**, 285–303.

Alberts, E.E., Schuman, G.E. and Burwell, R.E. (1978) Seasonal losses of
nitrogen and phosphorus from Missouri valley loess watersheds. *Journal
of Environmental Quality*, **7**, 203–8.

Allan, R.J. (1980) The inadequacy of existing chlorophyll 'a'/phosphorus
concentration correlations for assessing remedial measures for hypertrophic lakes. *Environmental Pollution* (Series B), **1**, 217–31.

Allen, K.R. (1951) The Horokiwi stream. A study of a trout population.
Fisheries Bulletin of the New Zealand Marine Department, **10**, 1–238.

Allen, H.E. and Kramer, J.R. (1972) *Nutrients in Natural Waters*. Wiley
Interscience, New York.

Allinson, A. and Newton, I. (1974) Waterfowl at Loch Leven, Kinross. *Proceedings of the Royal Society of Edinburgh*, B, **74**, 365–81.

Amemiya, M. (1970) Land and water management for minimising sediment, in *Agricultural Practices and Water Pollution* (eds.T. Willrich and G.E. Smith), Iowa University Press, Iowa, pp. 35–45.

Andersen, F.O. and Jensen, H.S. (in press) The influence of chironomids on decomposition of organic matter and nutrient exchange in a lake sediment. *Verhandlungen Internationale Vereinigung für Theoretische und Angewandte Limnologie*, **24**.

Anderson, J.M. (1971) Nitrogen and phosphorus budgets and the role of sediments in six shallow Danish lakes. *Archiv für Hydrobiologie*, **74**, 528–50.

Anderson, J.M. (1982) Effect of nitrate concentration in lake water on phosphate release from the sediment. *Water Research*, **16**, 1119–26.

Andersson, G., Cronberg, G. and Gelin, C. (1973) Planktonic changes following the restoration of Lake Trummen, Sweden. *Ambio*, **2**, 44–8.

Andersson, G., Berggren, H., Gronberg, G. and Gelin, C.I. (1978) Effects of planktonic fish on organisms and water chemistry in eutrophic lakes. *Hydrobiologia*, **59**, 9–15.

Anon. (1949) Report of algal growths research group. *Journal of the Institution of Water Engineers*, **3**, 329–36.

Anon. (1971) *Algal Assay Procedure Bottle Test*. United States Environmental Protection Agency, Washington.

Anon. (1974) *The Loch Leven IBP Project*. Royal Society of Edinburgh, Edinburgh.

Anon. (1980) *Strategy for the UK Forest Industry*. Centre for Agricultural Strategy, Reading, UK.

Anon. (1982a) *Eutrophication of Waters. Monitoring, Assessment and Control*. Organisation for Economic Cooperation and Development, Paris.

Anon. (1982b) *Towards a Nature Conservation Strategy for Broadland*. Report of the Ecology Working Group, Broads Authority, Norwich.

Anon. (1983) *The Nitrogen Cycle of the United Kingdom*. The Royal Society, London.

Anon. (1986) *Water Pollution by Fertilisers and Pesticides*. Organisation for Economic Cooperation and Development, Paris.

Anon. (1987) *Broads Strategy and Management Plan*. Broads Authority, Norwich.

Anon. (1990) *Toxic Blue Green Algae*. National Rivers Authority, Kingfisher House, Orton Goldhay, Peterborough PE2 0ZR, England.

Antonie, R.L. (1978) Nitrogen control with the rotating biological contractor, in *Advances in Water and Wastewater Treatment*. (eds. W.W. Eckenfelder and M.P. Wanielista), Ann Arbor Science, Ann Arbor, Michigan.

Appleton, T. (1982) Rutland Water Nature Reserve: concept, design and management. *Hydrobiologia*, **88**, 211–24.

Arnold, D.E. (1971) Ingestion, assimilation, survival and reproduction by *Daphnia pulex* fed seven species of blue-green algae. *Limnology and Oceanography*, **16**, 906–20.

Asbirk, S. and Dybbro, T. (1978) Population size and habitat selection of the great-crested grebe (*Podiceps cristatus*) in Denmark 1975. *Dansk Ornithologisk Forening Tidsskrift*, **72**, 1–14.

Ashley, K.I. (1985) Hypolimnion aeration: practical design and application. *Water Research*, **6**, 735–40.

Ayles, G.B., Lark, J.G., Barica, J. and Kling, H. (1976) Seasonal mortality of rainbow trout (*Salmo gairdneri*) planted in small eutrophic lakes of central Canada. *Journal of the Fisheries Research Board of Canada*, **33**, 647–55.

Azam, F., Fenchel, T., Field, J.G., Gray, J.S., Meyer-Reil, L.A. and Thingstad, F. (1983) The ecological role of water column microbes in the sea. *Marine Ecology Progress Series*, **10**, 257–63.

Bagenal, T. (1978) *Methods for Assessment of Fish Production in Freshwaters*. Blackwells Scientific Publications, Oxford.

Bailey-Watts, A.E. and Kirika, A. (1987) A re-assessment of phosphorus inputs to Loch Leven (Kinross, Scotland): rationale and an overview of results on instantaneous loadings with special reference to runoff. *Transactions of the Royal Society of Edinburgh: Earth Sciences*, **78**, 351–67.

Bailey-Watts, A.E. and Maitland, P.S. (1984) Eutrophication and fisheries in Loch Leven, Kinross, Scotland. *Proceedings of the Conference of the Institute of Fisheries Management*, 170–90.

Balls, H., Moss, B. and Irvine, K. (1989) The loss of submerged plants with eutrophication 1. Experimental design, water chemistry, aquatic plant and phytoplankton biomass in experiments carried out in ponds in the Norfolk Broadland. *Freshwater Biology*, **22**, 71–87.

Balmér, P. and Hultman, B. (1988) Control of phosphorus discharges: present situation and trends. *Hydrobiologia*, **170**, 305–19.

Barham, P.J. (1986) *The Use of the Secondary Draw-Off Tower at Rutland Water*. Unpublished Report, Anglian Water, Oundle, Peterborough.

Barica, J. (1974) Some observations on internal recycling, regeneration and oscillation of dissolved nitrogen and phosphorus in shallow, self-contained lakes. *Archiv für Hydrobiologie*, **73**, 334–60.

Barica, J. (1978) Collapses of Aphanizomenon flos-aquae blooms resulting in massive fish kills in eutrophic lakes. *Verhandlungen Internationale Vereinigung Für Theoretische und Angewandte Limnologie*, **20**, 208–13.

Barica, J. and Armstrong, F.A.J. (1971) Contribution by snow to the nutrient budget of some small Northwest Ontario lakes. *Limnology and Oceanography*, **16**, 891–99.

Barica, J. and Mur, L.R. (1980) *Hypereutrophic Ecosystems*. Dr. W. Junk BV, The Hague.

Barko, J.W. and Smart, R.M. (1981) Sediment-based nutrition of submerged macrophytes. *Aquatic Botany*, **10**, 229–38.

Barrett, P.R.F. (1978) Aquatic weed control: necessity and methods. *Fisheries Management*, **9**, 93–101.

Barsdate, R.J., Prentki, R.T. and Fenchel, T. (1974) Phosphorus cycle of model ecosystems: significance for decomposer food chains and effect of bacterial grazers. *Oikos*, **25**, 239–51.

Bayley, R.W. (1971) Nutrient removal from wastewaters. *Institution of Public Health Engineers Journal*, **70**, 150–72.

Bays, J.S. and Crisman, T.L. (1983) Zooplankton and trophic state relationships in Florida lakes. *Canadian Journal of Fisheries and Aquatic Sciences*, **40**, 1813–19.

Beaulac, M.N. (1980) *Sampling design and nutrient export coefficients: an examination of variability within differing land uses.* M.Sc., Michigan State University, East Lancing.

Beeton, A.M. (1965) Eutrophication of the St. Lawrence Great Lakes. *Limnology and Oceanography*, **10**, 240–54.

Beeton, A.M. (1969) Changes in the environment and biota of the Great Lakes, in *Eutrophication: Causes, Consequences, Correctives* (ed . G.A. Rohlich), National Academy of Sciences, Washington, DC, pp. 150–87.

Belcher, H. and Swale, E. (1979) *An Illustrated Guide to River Phytoplankton.* Her Majesty's Stationary Office, London.

Bengtsson, L. and Gelin, C. (1975) Artificial aeration and suction dredging methods for controlling water quality, in *Effects of Storage on Water Quality*, Water Research Centre, Medmenham, pp. 314–42.

Benndorf, J. (1979) A contribution to the phosphorus loading concept. *Internationale Revue Gesampten Hydrobiologie*, **64**, 177–88.

Benndorf, J. (1987) Food web manipulation without nutrient control: a useful strategy in lake restoration? *Schweizerische Zeitschrift fuer Hydrologie*, **49**, 237–48.

Benndorf, J. and Horn, W. (1985) Theoretical considerations on the relative importance of food limitation and predation in structuring zooplankton communities. *Archiv für Hydrobiologie Ergebnisse der Limnologie*, **21**, 383–96.

Benndorf, J. and Recknagel, F. (1982) Problems of application of the ecological model SALMO to lakes and reservoirs having various trophic states. *Ecological Modelling*, **17**, 129–45.

Benzie, W.J. and Courchaine, R.J. (1966) Discharges from separate storm sewers and combined sewers. *Journal of the Water Pollution Control Federation*, **38**, 410–21.

Berg, N.A. (1979) Control of phosphorus from agricultural land in the Great Lakes basin, in Phosphorus Management Strategies for Lakes, (eds. R.C. Loehr, C.S. Martin and W. Rast), Ann Arbor Science, Ann Arbor, pp. 459–86.

de Bernardi, R. and Giussani, G. (1990) Are blue-green algae a suitable food for zooplankton? An overview. *Hydrobiologia*, **200/201**, 29–41.

Bernhardt, H. (1977) Ten years' experience of reservoir aeration. *Progress in Water Technology*, **7**, 483–95.

Best, M.D. and Mantai, K.E. (1978) Growth of *Myriophyllum*: sediment or lake water as the source of nitrogen and phosphorus. *Ecology*, **59**, 1075–80.

Bird, D.F. and Kalff, J. (1984) Empirical relationships between bacterial abundance and chlorophyll concentration in fresh and marine waters. *Canadian Journal of Fisheries and Aquatic Sciences*, **41**, 1015–23.

Bird, D.F. and Kalff, J. (1986) Bacterial grazing by planktonic lake algae. *Science*, **231**, 493–5.

Bloesch, J. and Burns, N.M. (1980) A critical review of sediment trap techniques. *Schweizerische Zeitschrift fuer Hydrologie*, **42**, 15–55.

Bolas, P. and Lund, J.W.G., (1974) Some factors affecting the growth of *Cladophora glomerata* in the Kentish Stour. *Water Treatment and Examination*, **23**, 25–51.

Bolin, B. and Arrhenius, E. (1977) Nitrogen – an essential life factor and a growing environmental hazard. *Ambio*, **6**, 96–181.

Borman, F.H. and Likens, G.E. (1967) Nutrient cycling. *Science*, **155**, 424–9.

Bormann, F.H., Likens, G.E., Fisher, D.W. and Pierce, R.S. (1968) Nutrient loss accelerated by clear-cutting of a forest ecosystem. *Science*, **159**, 882–4.

Bormann, F.H., Likens, G.E., Siccama, T.G., Pierce, R.S. and Eaton, J.S. (1974) The export of nutrients and recovery of stable conditions following deforestation at Hubbard Brook. *Ecological Monographs*, **44**, 255–77.

Boström, B., Ahlgren, I. and Bell, R. (1985) Internal nutrient loading in an eutrophic lake, reflected in seasonal variations of some sediment parameters. *Verhandlungen Internationale Vereinigung für Theoretische und Angewandte Limnologie*, **22**, 3335–9.

Boström, B., Andersen, J.M., Fleischer, S. and Jansson, M. (1988) Exchange of phosphorus across the sediment-water interface. *Hydrobiologia*, **170**, 229–44.

Braband, Å., Faafeng, B. and Nilssen, J.P. (1986) Juvenile roach and invertebrate predators: delaying the recovery phase of eutrophic lakes by suppression of efficient filter-feeders. *Journal of Fish Biology*, **29**, 99–106.

Brezonik, P.L. (1972) Nitrogen: sources and transformations in natural waters, in *Nutrients in Natural Waters* (eds. H.E. Allen and J.R. Kramer), Wiley Interscience, New York, pp. 1–50.

Brierley, S.J. (1984) *The Effects of Artificial Overturn on Algal Populations in Reservoirs*. Severn-Trent Water Authority, Birmingham, report number RP–85–070.

Brierley, S.J. (1985) *The effects of artificial mixing on phytoplankton growth and periodicity*. Ph.D. thesis, University of Leicester.

Brierley, S.J., Harper, D.M. and Barham, P.J. (1989) Factors affecting the distribution and abundance of aquatic plants in a navigable lowland river; the River Nene, England. *Regulated Rivers: Research and Management*, **4**, 263–74.

Brink, N. (1975) Water pollution from agriculture. *Journal of the Water Pollution Control Federation*, **47**, 789–95.

Brinson, M.M., Bradshaw, D. and Holmes, R.N. (1983) Significance of floodplain sediments in nutrient exchange between a stream and its floodplain, in *Dynamics of Lotic Ecosystems* (eds. T.D. Fontaine III and S.M. Bartell), Ann Arbor Science, Ann Arbor, pp. 199–221.

Brockson, R.W., Davies, G.E. and Warren, C.E. (1970) Analysis of the trophic processes on the basis of density-dependent functions, in *Marine Food Chains* (ed. J.H. Steele), University of California Press, Berkeley, pp. 468–98.

Brook, A.J. (1959) The status of desmids in the plankton and the determination of phytoplankton quotients. *Journal of Ecology*, **47**, 429–45.

Brook, A.J. (1964) The phytoplankton of the Scottish freshwater lochs, in *The Vegetation of Scotland* (ed. A.J. Burnett), Oliver and Boyd, Edinburgh, pp. 290–305.

Brook, A.J. (1965) Planktonic algae as indicators of lake types, with special reference to the Desmidaceae. *Limnology and Oceanography*, **10**, 403–11.

Brooker, M.P., Morris, D.L. and Hemsworth, R.J. (1977) Mass mortalities of adult salmon (*Salmo salar*) in the River Wye, 1976. *Journal of Applied Ecology*, **14**, 409–17.

Brookes, A. (1988) *Channelised Rivers: Perspectives for Environmental Management*, John Wiley and Sons Ltd., Chichester.

Brooks, J.L. (1969) Eutrophication and changes in the composition of the zooplankton, in Eutrophication: Causes, Consequences, Correctives (ed. G.A. Rohlich), National Academy of Sciences, Washington, DC, pp. 236-55.

Brooks, J.L. and Dodson, S.I. (1965) Predation, body size and composition of plankton. *Science*, **150**, 28–35.

Brylinsky, M. (1980) Estimating the productivity of lakes and reservoirs, in *The Functioning of Freshwater Ecosystems* (eds. E.D. Le Cren and R.H. Lowe-McConnell), Cambridge University Press, Cambridge, pp. 411–54.

Brylinsky, M. and Mann, K.H. (1973) An analysis of factors governing productivity in lakes and reservoirs. *Limnology and Oceanography*, **18**, 1–14.

Bull, C.J. and Mackay, W.C. (1976) Nitrogen and Phosphorus removal from lakes by fish harvest. *Journal of the Fisheries Research Board of Canada*, **33**, 1374–6.

Burns, C.W. (1968) The relationship between body size of filter-feeding cladocera and the maximum size of particle ingested. *Limnology and Oceanography*, **13**, 675–8.

Burns, C.W. (1969) Relation between filtering rate, temperature and body size in four species of Daphnia. *Limnology and Oceanography*, **14**, 693–400.

Burton, T.M., King, D.L. and Ervin, J.L. (1979) Aquatic plant harvesting as a lake restoration technique, in *Lake Restoration: Proceedings of a National Conference*, United States Environmental Protection Agency, Mineapolis, Minnesota, pp. 177–84.

Canfield, D.E. (1983) Prediction of chlorophyll 'a' concentrations in Florida lakes: the importance of phosphorus and nitrogen. *Water Resources Bulletin*, **19**, 255–62.

Canfield, D.E. and Bachmann, R.W. (1981) Prediction of total phosphorus concentrations, chlorophyll a, and Secchi depths in natural and artificial lakes. *Canadian Journal of Fisheries and Aquatic Science*, **38**, 414–23.

Canfield, D.E., Shireman, J.V., Colle, D.E., Haller, W.T., Watkins, C.E. and Maceina, M.J. (1984) Prediction of chlorophyll 'a' concentrations in Florida lakes: importance of aquatic macrophytes. *Canadian Journal of Fisheries and Aquatic Science*, **41**, 497–501.

Carlson, R.E. (1977) A trophic state index for lakes. *Limnology and Oceanography*, **22**, 361–9.

Carpenter, S.R., Kitchell, J.K. and Hodgson, J.R. (1985) Cascading trophic interactions and lake productivity. *BioScience*, **35**, 634–9.

Chapman, V.J., Brown, J.M.A., Hill, C.F. and Carr, J.L. (1974) Biology of excessive weed growth in the hydro-electric lakes of the Waikato River, New Zealand. *Hydrobiologia*, **44**, 349–63.

Chapra, S.C. (1975) Comment on 'An empirical method of estimating the retention of phosphorus in lakes' by W.B. Kirchner and P.J. Dillon. *Water Resources Research*, **11**, 1033–4.

Chapra, S.C. (1977) Total phosphorus model for the Great Lakes. *Journal of the Environmental Engineering Division, American Society for Civil Engineering*, **103**, 147–61.

Chen, C.W. and Orlob, G.T. (1975) Ecologic simulation for aquatic environments, in *Systems Analysis and Simulation in Ecology* (ed. B.C. Patten), Academic Press, New York, pp. 475–588.

Chen, C.W. and Smith, D.J. (1979) Preliminary insights into a three-dimensional ecological-hydrodynamic model, in *Perspectives on Lake Ecosystem Modelling* (eds. D. Scavia and A. Robertson), Ann Arbor Science, Ann Arbor, pp. 249–79.

Chiaudani, G. and Vighi, M. (1974) The N:P ratio and tests with Selenastrum to predict eutrophication in lakes. *Water Research*, **8**, 1063–9.

Chiaudani, G., Premazzi, G. and Vighi, M. (1986) The impact of the River Po basin land use on the eutrophication of northern Adriatic coastal waters, in *Effects of Land Use on Fresh Waters* (ed. J.F.G. Solbé), Ellis Horwood, Chichester, pp. 488–92.

Coesel, P.F.M. (1975) The relevance of desmids in the biological typology and evaluation of freshwaters. *Hydrobiological Bulletin*, **9**, 93–101.

Colby, P.J., Spangler, G.R., Hurley, D.A. and McCombie, A.M. (1972)

Effects of eutrophication on salmonid communities in oligotrophic lakes. *Journal of the Fisheries Research Board of Canada*, **29**, 975–83.

Collie, A. and Lund, J.W.G. (1981) Bioassays of water from the Severn and Thames river systems. *Journal of the Institution of Water Engineers and Scientists*, **34**, 180–7.

Collingwood, R.W. (1977) *A survey of eutrophication in Britain and its effects upon water supplies*. Water Research Centre, Medmenham, Technical Report 40.

Collingwood, R.W. (1978) The dissipation of phosphorus in sewage and sewage effluent, in *Phosphorus in the environment: its chemistry and biochemistry*. Elsevier, Amsterdam, pp. 229–42.

Connor, S. (1985) River research gets washed away. *New Scientist*, **107**, 38–40.

Cooke, G.W. (1975) A review of the effects of agriculture on the chemical composition and quality of surface and underground waters. *Ministry of Agriculture, Fisheries and Food Technical Bulletin*, **32**, 5–57.

Cooke, G.W. and Williams, R.J.B. (1970) Losses of nitrogen and phosphorus from agricultural land. *Water Treatment and Examination*, **19**, 253–76.

Cooke, G.W. and Williams, R.J.B. (1973) Significance of man-made sources of phosphorus: fertilisers and farming. *Water Research*, **7**, 19–33.

Cooke, C.D., Welch, E.B., Peterson, S.A. and Newroth, P.R. (1986) *Lake and Reservoir Restoration*. Butterworths, London.

Cornett, R.J. and Rigler, F.H. (1979) Hypolimnetic oxygen deficits: their prediction and interpretation. *Science*, **205**, 580–1.

Cowell, B.C., Dawes, C.J., Gardiner, W.E. and Scheda, S.M. (1987) The influence of whole lake aeration on the limnology of a hypereutrophic lake in central Florida. *Hydrobiologia*, **148**, 3–24.

Craig, J.F. and Kipling, C. (1983) Reproductive effort versus the environment; case histories of Windermere perch, *Perca fluviatilis* L. and pike, *Esox lucius* L. *Journal of Fish Biology*, **22**, 713–27.

Crisman, T.L. and Beaver, J.R. (1990) Applicability of planktonic biomanipulation for managing eutrophication in the subtropics. *Hydrobiologia*, **200/201**, 177–85.

Crisp, D.T. and Mann, R.H.K. (1977) A desk study of the performance of trout fisheries in a selection of British reservoirs. *Fisheries Management*, **8**, 101–19.

Crook, C., Boar, R. and Moss, B. (1983) *The Decline of Reedswamp in the Norfolk Broadland: Causes, Consequences and Solutions*. Research Series Report 6, Broads Authority, Norwich, England.

Crowder, A.A., McLaughlin, B., Weir, R.D. and Christie, W.J. (1986) Shoreline fauna of the bay of Quinte. *Canadian Special Publications of Fisheries and Aquatic Sciences*, **86**, 190–200.

Cryer, M., Pierson, G. and Townsend, C. (1986) Reciprocal interactions between roach, *Rutilus rutilus* and zooplankton in a small lake: prey dynamics and fish growth and recruitment. *Limnology and Ocean-ography*, **31**, 1022–38.

Cullen, P. (1986) Managing nutrients in aquatic systems: the eutrophication problem. In *Limnology in Australia* (eds. P. De Deckker and W.D. Williams), CSIRO, Melbourne, Australia, pp. 539–54.

Cullen, P. and Fosberg, C. (1988) Experiences with reducing point sources of phosphorus to lakes. *Hydrobiologia*, **170**, 321–36.

Cullen, P., Farmer, N. and O'Loughlin, E. (1988) Estimating non-point sources of phosphorus in lakes. *Verhandlungen Internationale Vereinigung für Theoretische und Angewandte Limnologie*, **23**, 588–93.

Currie, D.J. and Kalff, J. (1984) A comparison of the abilities of bacteria and algae to acquire and retain phosphorus. *Limnology and Ocean-ography*, **29**, 298–310.

Deevey, E.S. (1941) Limnological Studies in Connecticut. VI. The quantity and composition of the bottom fauna of thirty-six Connecticut and New York lakes. *Ecological Monographs*, **11**, 413–55.

Deevey, E.S. (1942) Studies on Connecticut lake sediments III. The biostratification of Linsley Pond. *American Journal of Science*, **240**, 233–64, 313–24.

Deevey, D.G. and Harkness, N. (1973) The significance of man-made sources of phosphorus: detergents and sewage. *Water Research*, **7**, 35–54.

Dejoux, C. (1988) *La Pollution des Eaux Continentales Africaines. Expérience acquise Situation Actuelle et Perspectives.* ORSTOM Collection Travaux et Documents, Paris.

Dillon, P.J. (1974a) *The Application of the Phosphorus Loading Concept to Eutrophication Research*, Report No. 7, National Research Council of Canada.

Dillon, P.J. (1974b) A critical review of Vollenweider's nutrient budget model and other related models. *Water Resources Bulletin*, **10**, 969–89.

Dillon, P.J. and Kirchner, W.B. (1974) The effects of geology and land use on the export of phosphorus from watersheds. *Water Research*, **9**, 135–48.

Dillon, P.J. and Kirchner, W.B. (1975) Reply. *Water Resources Research*, **11**, 1035–6.

Dillon, P.J. and Rigler, F.H. (1974a) A test of a simple nutrient budget model predicting the phosphorus concentration in lake water. *Journal of the Fisheries Research Board of Canada*, **31**, 1771–8.

Dillon, P.J. and Rigler, F.H. (1974b) The phosphorus-chlorophyll relationship in lakes. *Limnology and Oceanography*, **19**, 767–73.

Dillon, P.J. and Rigler, F.H. (1975) A simple method for predicting the capacity of a lake for development based upon lake trophic status. *Journal of the Fisheries Research Board of Canada*, **32**, 1519–31.

van Donk, E., Gulati, R.D. and Grimm, M.P. (1989) Food web manipulation in Lake Zwemlust: positive and negative effects during the first two years. *Hydrobiological Bulletin*, **23**, 19–34.

van Donk, E., Grim, M.P., Gulati, R.D. and Breteler, J.P.G.K. (1990) Whole-lake food-web manipulation as a means to study community interactions in a small ecosystem. *Hydrobiologia*, **200/201**, 275–89.

van Donk, E., Grimm, M.P., Gulati, R.D., Heuts, P.G.M., de Kloet, A.W. and van Liere, L. (1990) First attempt to apply whole-lake food-web manipulation on a large scale in The Netherlands. *Hydrobiologia*, **200/201**, 291–301.

Doudoroff, P. (1970) *Dissolved oxygen requirements of freshwater fishes*, Fisheries technical Paper 86, FAO, Rome.

Duarte, C.M. and Kalff, J. (1986) Littoral slope as a predictor of the maximum biomass of submerged macrophyte communities. *Limnology and Oceanography*, **31**, 1072–80.

Duarte, C.M., Kalff, J. and Peters, R.H. (1985) Patterns in biomass and cover of aquatic macrophytes in lakes. *Canadian Journal of Fisheries and Aquatic Science*, **43**, 1900–8.

Duncan, A. (1985) Body carbon in daphnids as an indicator of the food concentration available in the field. *Archiv für Hydrobiologie Beiheft. Ergebnisse der Limnologie*, **21**, 81–90.

Eckenfelder, W.W. (1969) Development of tertiary treatment methods for wastewater renovation. *Water Pollution Control*, **68**, 584–91.

Edington, J.M. and Edington, M.A. (1977) *Ecology and Environmental Planning*. Chapman and Hall Ltd, London.

Edington, J.M. and Edington, M.A. (1986) Ecology, Recreation and Tourism. Cambridge University Press, Cambridge.

Edmonson, W.T. and Lehman, J.T. (1981) The effects of changes in the nutrient income on the condition of Lake Washington. *Limnology and Oceanography*, **26**, 1–29.

Edwards, R.W. and Brooker, M.P. (1982) *The Ecology of the Wye*. Dr. W. Junk Publishers, The Hague.

Ellis, B.G. and Childs, K.E. (1973) *Nutrient movement from septic tanks and lawn fertilisation*. Technical Bulletin 73–5, Dept. of Natural Resources, Lancing, Michigan.

Eltringham, S.K. (1984) *Wildlife Resources and Economic Development*. John Wiley and Sons Ltd., Chichester.

Eminson, D. and Phillips, G.L. (1978) A laboratory experiment to examine the effects of nutrient enrichment on macrophyte and epiphyte growth. *Verhandlungen Internationale Vereinigung für Theoretische und Angewandte Limnologie*, **20**, 82–7.

European-Economic-Community. (1980) Council directive on the quality of water for human consumption. *Official Journal*, **23** (80/778 EEC L 229), 11–29.

Famme, P. and Knudsen, J. (1985) Anoxic survival, growth and reproduction by the freshwater annelid Tubifes sp. demonstrated using a new simple anoxic chemostat. *Comparative Biochemistry and Physiology*, **81A**, 251–3.

Fast, A. (1979) Artificial aeration as a lake restoration technique, in *Lake*

Restoration: Proceedings of a National Conference, United States Environmental Protection Agency, Washington, DC, EPA 440/5–79–001, pp. 121–5.

Fast, A.W. (1981) The effects of artificial destratification on algal poulations, in *Destratification of Lakes and Reservoirs to Improve Water Quality* (ed. Burns, F.L. and Powling, I.J.), Australian Water Resources Council, Canberra, pp. 515–56.

Ferguson, A.J.D. and Harper, D.M. (1982) Rutland Water phytoplankton: the development of an asset or a nuisance? *Hydrobiologia*, **88**, 117–33.

Fiala, L. and Vasata, P. (1982) Phosphorus reduction in a man-made lake by means of a small reservoir on the inflow. *Archiv für Hydrobiologie*, **94**, 24–37.

Flade, M. (1979) Large increase in population of the great crested grebe Podiceps cristatus, in the area of Wolfsburg, West Germany, and causative factors. *Vogelkundliche Berichte aus Niedersachsen*, **11**, 33–40.

Flett, R.J., Schindler, D.W., Hamilton, R.D. and Campbell, N.E.R. (1980) Nitrogen fixation in Canadian Precambrian Shield lakes. *Canadian Journal of Fisheries and Aquatic Sciences*, **37**, 494–505.

Fogg, G.E. (1975) *Algal Cultures and Phytoplankton Ecology*. University of London Press, London.

Fogg, G.E. (1980) Primary Production. In *Fundamentals of Aquatic Ecosystems* (eds. Barnes, R.S.K. and Mann, K.H.), Blackwell Scientific Publications, Oxford, pp. 24–45.

Fogg, G.E. (1983) The ecological significance of extracellular products of phytoplankton photosynthesis. *Bot. Mar.*, **26**, 3–14.

Ford, D.B., Drage, B.E. and Roberts, T.J. (1982) The production of public supply drinking water from Rutland Water at the Wing Treatment Works. *Hydrobiologia*, **88**, 103–16.

Fosberg, B.R. and Shapiro, J. (1980) Predicting the algal response to destratification, in *Restoration of Lakes and Inland Waters*, US Environmental Protection Agency, Washington, DC, EPA 440/5–81–010, pp. 134–9.

Fosberg, C. (1980) Detergent modification: Scandanavian experiences, in *Restoration of Lakes and Inland Waters*, US Environmental Protection Agency, Washington, DC, EPA 440/5–81–010, pp. 429–31.

Fosberg, C. and Ryding, S. (1980) Eutrophication parameters and trophic state indices in 30 Swedish waste-receiving lakes. *Archiv für Hydrobiologie*, **89**, 189–207.

Fosberg, C., Ryding, S., Claesson, A. and Fosberg, A. (1978) Water chemical analyses and/or algal assay? – sewage effluent and polluted water studies. *Mitteilungen Internationale Vereinigung für Theoretische und Angewandte Limnologie*, **21**, 352–63.

Foy, R.H. (1985) Phosphorus inactivation in a eutrophic lake by the direct addition of ferric aluminium sulphate: impact on iron and phosphorus. *Freshwater Biology*, **15**, 613–29.

Froelich, P.N. (1988) Kinetic control of dissolved phosphate in natural rivers and estuaries: a primer on the phosphate buffer mechanism. *Limnology and Oceanography*, 33, 649–68.

Frost, B.W. (1980) Grazing, in *The Physiological Ecology of Phytoplankton* (ed. I. Morris), Blackwell Scientific Publishers, Oxford, pp. 465–91.

Fryer, G. (1987) Quantitative and qualitative: numbers and reality in the study of living organisms. *Freshwater Biology*, 17, 177–89.

Gachter, R. and Furrer, O.J. (1972) Der Beitrag der Landwirtschaft zur Eutrophierung der Gewasser in der Schweiz. 1. Ergebnisse von direkten Messungen im Einzugsgebiet verschiedener Vorfluter. *Schweizerische Zeitschrift fuer Hydrologie*, 34, 41–70.

Gascon, D. and Leggett, W.C. (1977) Distribution, abundance and resource utilisation of littoral zone fishes in response to a nutrient/production gradient in Lake Mempremagog. *Journal of the Fisheries Research Board of Canada*, 34, 1105–17.

Gauch, H.G.J. (1982) *Multivariate Analysis in Community Ecology.* Cambridge University Press, Cambridge.

Gaunlett, R.B. (1980) Removal of nitrogen compounds, in *Developments in Water Treatment* (ed. W.M. Lewis), Applied Science Publishers Ltd, London, pp. 59–88.

Geller, W. (1985) Production, food utilisation and losses of two coexisting, ecologically different Daphnia species. *Archiv für Hydrobiologie Beiheft. Ergebnisse der Limnologie*, 21, 67–79.

Geller, W. and Müller, H. (1981) The filtration apparatus of cladocera: filter mesh sizes and their implications on food selectivity. *Oecologia* (Berlin), 49, 316–21.

George, D.G., Hewitt, D.P., Lund, J.W. G. and Smyly, W.J.P. (1990) The relative effects of enrichment and climatic change on the long-term dynamics of *Daphnia* in Estwaite Water, Cumbria. *Freshwater Biology*, 23, 55–70.

George, M. (1970) *The Aquatic Macrophytes of the Broads.* Nature Conservancy Council, Norwich.

George, M. (1976) Land use and nature conservation in Broadland. *Geography*, 67, 131–42.

George, M. (in press) *The Land Use, Ecology and Conservation of Broadland.* Packard Publishing, Chichester.

Gerking, S.D. (1978) *Ecology of Freshwater Fish Production.* John Wiley and Sons, New York.

Gerloff, G.C. (1969) Evaluating nutrient supplies for the growth of aquatic plants in natural waters, in *Eutrophication: Causes, Consequences, Correctives* (ed. G.A. Rohlich), National Academy of Sciences, Washington, D.C, pp. 537–55.

Gersberg, R.M., Elkins, B.W. and Goldman, C.R. (1983) Nitrogen removal in artificial wetlands. *Water Research*, 17, 1009–14.

Gliwicz, Z.M. (1969) Studies on the feeding of pelagic zooplankton in lakes with varying trophy. *Ekologia Polska*, 17, 663–708.

Gliwicz, Z.M. (1975) Effect of zooplankton grazing on photosynthetic activity and composition of zooplankton. *Verhandlungen Internationale Vereinigung für Theoretische und Angewandte Limnologie*, 19, 1490–7.

Gliwicz, Z.M. (1977) Food size selection and seasonal succession of filter feeding zooplankton in an eutrophic lake. *Ekologia Polska*, 25, 179–225.

Gliwicz, Z.M. (1978) Filtering rates, food size selection, and feeding rates in cladocerans – another aspect of interspecific competition in filter-feeding zooplankton, in *Evolution and Ecology of Zooplankton Communities* (ed. W.C. Kerfoot), University Press of New England, Hanover, New Hampshire, pp. 282–98.

Gliwicz, Z.M. (1985) Predation or food limitation: an ultimate reason for extinction of planktonic cladoceran species. *Archiv für Hydrobiologie. Ergebnisse der Limnologie*, 21, 419–30.

Gliwicz, Z.M. (1990) Why do cladocerans fail to control algal blooms? *Hydrobiologia*, 200/201, 83–97.

Goldman, C.R. (1960) Molybdenum as a factor limiting primary production in Castle Lake, California. *Science*, 132, 1016–7.

Goldman, C.R. (1964) Primary productivity and micronutrient limiting factors in some North American and New Zealand lakes. *Verhandlungen Internationale Vereinigung für Theoretische und Angewandte Limnologie*, 15, 365–74.

Goldman, C.R. and Horne, A.J. (1983) *Limnology*. McGraw-Hill International Book Company, New York.

Goldman, C.R., Porcella, D.B., Middlebrooks, E.J. and Torein, D.F. (1972) The effect of carbon on algal growth – its relationship to eutrophication. *Water Research*, 6, 637–79.

Golterman, H.L. (1973) Natural phosphate sources in relation to phosphate budgets: a contribution to the understanding of eutrophication. *Water Research*, 7, 3–17.

Golterman, H.L. (1975) *Physiological Limnology*. Elsevier Scientific Publishing, Amsterdam.

Golterman, H.L. (1976) Zonation of mineralisation in stratifying lakes, in *The Role of Terrestrial and Aquatic Organisms in Decomposition Processes* (eds. J.M. Anderson and J. Macfadyen), Blackwell Scientific Publications, Oxford, pp. 3–32.

Golterman, H.L. (1984) Sediments, modifying and equilibrating factors in the chemistry of freshwater. *Verhandlungen Internationale Vereinigung für Theoretische und Angewandte Limnologie*, 22, 23–59.

Gophen, M. (1984) The impact of zooplankton status on the management of Lake Kinneret (Israel) *Hydrobiologia*, 113, 249–58.

Granall, U. and Lundgren, A. (1971) Nitrogen fixation in Lake Erken. *Limnology and Oceanography*, 16, 711–9.

Granéli, W. and Solander, D. (1988) Influence of aquatic macrophytes on phosphorus cycling in lakes. *Hydrobiologia*, 170, 245–66.

Greene, C.H. (1983) Selective predation in freshwater communities.

Internationale Revue den Gesampten Hydrobiologie, 68, 297–315.

Grimm, M.P. (1989) Northern pike (*Esox lucius* L.) and aquatic vegetation, tools in the management of fisheries and water quality in shallow waters. *Hydrobiological Bulletin*, 23, 59–66.

Grobler, D.C. (1985) Phosphorus budget models for simulating the fate of phosphorus in South African reservoirs. *Water South Africa*, 11, 219–30.

Grobler, D. C. and Silberbauer, M. J. (1985a) Eutrophication control: a look into the future. *Water South Africa*, 11, 69–78.

Grobler, D.C. and Silberbauer, M.J. (1985b) The combined effect of geology, phosphate sources and runoff on phosphate export from drainage basins. *Water Research*, 19, 975–81.

Haapanen, A. (1978) The summer behaviour and habitat use of the whooper swan, *Cignus cignus cignus*. *Riistatiet Julkaisuja*, 36, 49–82.

Hall, D.J. (1962) An experimental approach to the dynamics of a natural population of *Daphnia galeata mendotae*. *Ecology*, 45, 94–112.

Hall, D.J., Threlkeld, S.J., Burns, C.W. and Crowley, P.H. (1976) The size-efficiency hypothesis and the size structure of zooplankton communities. *Annual Review of Ecology and Systematics*, 7, 177–208.

Hall, G.H. and Jeffries, C. (1984) The contribution of nitrification in the water column and profundal sediments to the total oxygen deficit of the hypolimnion of a mesotrophic lake (Grasmere, English Lake District) *Microbial Ecology*, 10, 37–46.

Hamilton, R.D. (1972) The environmental acceptability of NTA: current research and areas of concern, in *Nutrients and Eutrophication: the Limiting Nutrient Controversy* (ed. G.E. Likens), American Society for Limnology and Oceanography, Lawrence, Kansas, pp. 217–28.

Hammer, M.J. (1986) *Water and Wastewater Technology* (2nd. ed.), J. Wiley and Sons, New York.

Haney, J.F. (1973) An in situ examination of grazing activities of natural zooplankton communities. *Archivs für Hydrobiologie*, 72, 87–132.

Hanley, P.K. and Murphy, M.D. (1973) Soil and fertiliser phosphorus in the Irish ecosystem. *Water Research*, 7, 197–210.

Hansford, R.G. and Ladle, M. (1979) The medical importance and behaviour of *Simulium austeni* Edwards (Diptera: Simuliidae) in England. *Bulletin of Entomological Research*, 69, 33–41.

Hanson, J.M. and Leggett, W.C. (1982) Empirical prediction of fish biomass and yield. *Canadian Journal of Fisheries and Aquatic Sciences*, 39, 257–63.

Hanson, J.M. and Peters, R.H. (1984) Empirical prediction of crustacean zooplankton biomass and profundal macrobenthos biomass in lakes. *Canadian Journal of Fisheries and Aquatic Sciences*, 41, 439–45.

Harper, D.M. (1978) *Limnological Studies on Three Scottish Lowland Freshwater Lochs*. Ph.D. thesis, University of Dundee.

Harper, D.M. (1986) The effects of artificial enrichment upon the planktonic and benthic communities in a mesotrophic to hypertrophic loch series in lowland Scotland. *Hydrobiologia*, **137**, 9–19.

Harper, D.M. and Bullock, J.A. (1982) *Rutland Water – Decade of Change*. Dr. W. Junk, Bv., The Hague.

Harper, D.M. and Stewart, W.D.P. (1987) The effects of land use upon water chemistry, particularly nutrient enrichment, in shallow lowland lakes: comparative studies of three lochs in Scotland. *Hydrobiologia*, **148**, 211–29.

Harriman, R. (1978) Nutrient leaching from fertilised forest watersheds in Scotland. *Journal of Applied Ecology*, **15**, 933–42.

Hartmann, J. (1977) Fischereiliche veranderungen in kulturbedingt eutrophierenden Seen. *Schweizerische Zeitscrift fuer Hydrologie*, **39**, 243–54.

Haslam, S.M. (1978) *River Plants*. Cambridge University Press, Cambridge.

Hasler, A.D. (1947) Eutrophication of lakes by domestic drainage. *Ecology*, **28**, 383–95.

Hawkes, H.A. (1983) Activated sludge, in *Ecological Aspects of Used Water Treatment* (eds. C.R. Curds and H.A. Hawkes), Academic Press, London.

Haworth, E.Y. (1969) The diatoms of a sediment core from Blea Tarn, Langdale. *Journal of Ecology*, **57**, 429–39.

Haworth, E.Y. (1979) The distribution of a species of *Stephanodiscus* in the recent sediments of Blelham Tarn, English Lake District. *Nova Hedwigia*, **64**, 395–409.

Haworth, E.Y. (1985) The highly nervous system of the English Lakes: aquatic ecosystem sensitivity to external changes, as demonstrated by diatoms. *Annual Report of the Freshwater Biological Association*, **53**, 60–79.

Haworth, E.Y. and Lund, J.W.G. (1984) *Lake Sediments and Environmental History*. Leicester University Press, Leicester.

Hayes, C.R. and Greene, L.A. (1984) The evaluation of eutrophication impact in public water supply reservoirs in East Anglia. *Water Pollution Control*, **83**, 42–51.

Healey, F.P. (1973) Inorganic nutrient uptake and deficiency in algae. *CRC Critical Reviews in Microbiology*, **3**, 69–113.

Healey, F.P. and Hendzel, L.L. (1976) Physiological changes during the course of blooms of *Aphanizomenon flos-aquae*. *Journal of the Fisheries Research Board of Canada*, **33**, 36–41.

Healey, F.P. and Hendzel, L.L. (1980) Physiological indicators of nutrient deficiency in lake phytoplankton. *Canadian Journal of Fisheries and Aquatic Sciences*, **37**, 442–53.

Hecky, R.E. and Kilham, P. (1988) Nutrient limitation of phytoplankton in freshwater and marine environments: a review of recent evidence on the effects of enrichment. *Limnology and Oceanography*, **33**, 796–822.

Henderson-Sellars, B. (1979) *Reservoirs*. Macmillan Press Ltd., London.

Hendersen-Sellars, B. and Markland, H.R. (1987) *Decaying Lakes: The Origin and Control of Cultural Eutrophication*. John Wiley and Sons, Chichester.

Henry, R., Tundisi, J.G. and Curi, P.R. (1984) Effects of phosphorus and nitrogen enrichment on the phytoplankton in a tropical reservoir (Lobo Reservoir, Brazil). *Hydrobiologia*, **118**, 177–85.

Hern, S.C., Lambou, V.W. and Williams, L.R. (1979) *Comparison of Models Predicting Ambient Lake Phosphorus Concentrations* U.S. Environmental Protection Agency, Washington, DC, US–EPA–600/3–79–012.

Hetling, L.J. and Sykes, R.M. (1973) Sources of nutrients in Candarago Lake. *Journal of the Water Pollution Control Federation*, **45**, 145–56.

Heyman, U. (1983) Relationship between production and biomass of phytoplankton in four Swedish lakes of different trophic status and humic content. *Hydrobiologia*, **101**, 89–104.

Heyman, U. and Lundgren, A. (1988) Phytoplankton biomass and production in relation to phosphorus. *Hydrobiologia*, **170**, 211–27.

Hillbricht-Ilkowska, A. (1977) Trophic relations and energy flow in pelagic plankton. *Polish Ecological Studies*, **3**, 3–98.

Hilton, J. and Phillips, G.L. (1982) The effect of boat activity on turbidity in a shallow broadland river. *Journal of Applied Ecology*, **19**, 143–50.

Ho, Y.B. (1979) Shoot development and production studies of *Phragmites australis* (Cav.) Trin ex Steudel in Scottish Lochs. *Hydrobiologia*, **64**, 215–22.

Hobbie, J.E. and Likens, G.E. (1973) Output of phosphorus, dissolved organic carbon and fine particulate carbon from Hubbard Brook watersheds. *Limnology and Oceanography*, **18**, 734–42.

Hodges, L. (1973) *Environmental Pollution*. Holt, Reinhart and Winston, New York.

Holah, J.T. and McIver, J.D. (1982) The distribution of micro-organisms through a bunded lagoon in Rutland Water. *Hydrobiologia*, **88**, 225–9.

Holden, A.V. (1975) The relative importance of agricultural fertilisers as a source of nitrogen and phosphorus in Loch Leven. *Ministry of Agriculture, Fisheries and Food Technical Bulletin* **32**, 306–14.

Holtan, H. (1980) The case of Lake Mjòsa. *Progress in Water Technology*, **12**, 103–20.

Holtan, H., Kamp-Nielsen, L. and Stuanes, A.O. (1988) Phosphorus in soil, water and sediment: an overview. *Hydrobiologia*, **170**, 19–34.

Horie, S. (1981) On the significance of paleolimnological studies of ancient lakes – Lake Biwa and other relict lakes. *Verhandlungen Internationale Vereinigung für Theoretische und Angewandte Limnologie*, **21**, 13–44.

Horn, W. (1985) Results regarding the food of the planktonic crustaceans *Daphnia hyalina* and *Eudaptomus gracilis*. *Internationale Revue gesampten Hydrobiologie*, **70**, 703–9.

Horn, W. (1985) Investigations into the food selectivity of the planktonic crustaceans *Daphnia hyalina*, *Eudiaptomus gracilis* and *Cyclops vicinus*. *Internationale Revue der Gesampten Hydrobiologie*, **70**, 603–12.

Horne, A.J. and Vine, A.B. (1971) Nitrogen fixation and its significance in tropical Lake George, Uganda. *Nature*, **232**, 417–8.

Hosper, S.H. (1989) Biomanipulation, new perspectives for the restoration of shallow, eutrophic lakes in the Netherlands. *Hydrobiological Bulletin*, **23**, 5–10.

Hosper, S.H. and Jagtman, E. (1990) Biomanipulation additional to nutrient control for restoration of shallow lakes in The Netherlands. *Hydrobiologia*, **200/201**, 523–34.

Houghton, G.U. (1972) Long term increases in planktonic growth in the Essex Stour. *Water Treatment and Examination*, **21**, 299–308.

Howard-Williams, C. (1985) Cycling and retention of nitrogen and phosphorus in wetlands: a theoretical and applied perspective. *Freshwater Biology*, **15**, 391–431.

Howarth, R.W., Marino, M. and Cole, J.J. (1988) Nitrogen fixation in freshwater, estuarine and marine ecosystems. *Limnology and Oceanography*, **33**, 688–701.

Howarth, R.W., Marino, R., Lane, J. and Cole, J.J. (1988) Nitrogen fixation in freshwater, estuarine and marine ecosystems 1. Rates and importance. *Limnology and Oceanography*, **33**, 669–87.

Howells, W.R. and Merriman, R. (1986) Pollution from agriculture in the area of the Welsh Water Authority, in *Effects of Land Use on Freshwaters Agriculture, Forestry, Mineral Exploitation, Urbanisation* (eds. J.F. de L.G. Solbé), Ellis Horwood, Chichester, England, pp. 267–82.

Hrbacek, J., Novatna-Dvorakova, M., Korinek, V. and Prochazkova, L. (1961) Demonstration of the effect of fish stock on the species composition of the zooplankton and the intensity of metabolism of the whole plankton association. *Verhandlungen Internationale Vereinigung für Theoretische und Angewandte Limnologie*, **14**, 192–5.

Huber, (1982) *A classification of Florida lakes*. Florida Department of Environmental Regulation Report No. ENV–05–82–1.

Huet, M. (1970) *Traite de Pisciculture* (4th ed.), Ch. de Wyngaert, Brussels.

Hughes, J.C. and Lund, J.W.G. (1962) The rate of growth of Asterionella formosa Hass. in relation to its ecology. *Archiv für Microbiologie*, **42**, 117–29.

Hughes, H.R., McColl, R.H.S. and Rawlence, D.J. (1978) Lake Ellesmere, Canterbury, New Zealand. A review of the lake and its catchment. *New Zealand Department of Scientific and Industrial Research Information Series*, **99**, 1–27.

Hurley, D.A. (1986) Fish populations of the Bay of Quinte, Lake Ontario, before and after phosphorus control. *Canadian Special Publications of Fisheries and Aquatic Sciences*, **86**, 201–14.

Hutchinson, G.E. (1944) Nitrogen in the biogeochemistry of the atmosphere. *American Scientist*, **32**, 178–95.

Hutchinson, G.E. (1957) *A Treatise on Limnology, volume 1*. John Wiley and Sons, New York.

Hutchinson, G.E. (1973) Eutrophication. *American Scientist*, **61**, 269–79.

Hutchinson, G.E. and Bowen, V.T. (1950) A quantitative radiochemical study of the phosphorus cycle in Linsley Pond. *Ecology*, **31**, 194–203.

Hutchinson, G.E. and Cowgill, U.M. (1970) The history of the lake; A synthesis. Ianula: An account of the history and development of the Lago di Monterosi, Latium, Italy. *Transactions of the American Philosophical Society*, **60**, 163–70.

Hynes, H.B.N. (1969) The enrichment of streams, in *Eutrophication: Causes, Consequences, Correctives* (ed. G.A. Rohlich), National Academy of Sciences, Washington, DC, pp. 188–96.

Imboden, D.M. (1974) Phosphorus model of lake eutrophication. *Limnology and Oceanography*, **19**, 297–304.

Imboden, D.M. (1975) Restoration of a Swiss lake by internal measures: can models explain reality? in *Lake Pollution and Recovery* (eds. R. Vismara, R. Marforio, V.Mezzanotte, and S. Cernuschi), European Water Pollution Control Association, Rome, pp. 91–102.

Imboden, D.M. and Gachter, R. (1978) A dynamic lake model for trophic state prediction. *Ecological Modelling*, **4**, 77–98.

Irvine, K., Moss, B. and Balls, H. (1989) The loss of submerged plants with eutrophication 2. Relationships between fish and zooplankton in a set of experimental ponds, and conclusions. *Freshwater Biology*, **22**, 89–107.

Jamieson, C.D. (1980) The predatory feeding of copepodid stages III to adult *Mesocyclops leukarti* (Claus), in *Evolution and Ecology of Zooplankton Communities* (ed. W.C. Kerfoot), University Press of New England, Hanover, New Hampshire, pp. 518–37.

Jansson, M. (1988) Phosphate uptake and utilisation by bacteria and algae. *Hydrobiologia*, **170**, 177–89.

Jansson, M., Olsson, H. and Pettersson, K. (1988) Phosphatases; origin, characteristics and function in lakes. *Hydrobiologia*, **170**, 157–75.

Jassby, A.D. and Goldman, C.R. (1974) Loss rates from a lake phytoplankton community. *Limnology and Oceanography*, **19**, 618–27.

Jewson, D.H., Rippey, B.H. and Gilmore, W.K. (1981) Loss rates from sedimentation, parasitism and grazing during the growth, nutrient limitation and dormancy of a diatom crop. *Limnology and Oceanography*, **26**, 1045–56.

Johnson, A.H., Bouldin, D.R., Goyette, E.A. and Hedges, A.H. (1976) Phosphorus loss by stream transport from a rural watershed: quantities, processes and sources. *Journal of Environmental Quality*, **5**, 148–57.

Johnson, M.G. and McNeil, O.C. (1986) Changes in abundance and species composition in benthic invertebrate communities in the Bay of Quinte, 1966–84. *Canadian Special Publications of Fisheries and Aquatic Sciences*, **86**, 177–89.

Johnston, W.R., Ittihadieh, F., Daum, R.M. and Pillsbury, A.F. (1965) Nitrogen and phosphorus in tile drainage effluent. *Proceedings of the Soil Science Society of America*, **29**, 287–9.

Jónasson, P. (1969) Bottom fauna and eutrophication, in *Eutrophication: Causes, Consequences, Correctives* (ed. G.A. Rohlich), National

Academy of Sciences, Washington DC, pp. 274–305.

Jónasson, P. (1975) Population ecology and production of benthic detritivores. *Verhandlungen Internationale Vereinigung für Theoretische und Angewandte Limnologie*, **19**, 1066–72.

Jónasson, P. (1978) Zoobenthos of lakes. *Verhandlungen Internationale Vereinigung für Theoretische und Angewandte Limnologie*, **20**, 13–37.

Jónasson, P. and Kristiansen, J. (1967) Primary and secondary production in Lake Esrom. Growth of *Chironomus anthracinus* in relation to seasonal cycles of phytoplankton and dissolved oxygen. *Internationale Revue der Gesampten Hydrobiologie*, **52**, 163–217.

Jones, J.G. (1979) Microbial nitrate reduction in freshwater sediments. *Journal of General Microbiology*, **115**, 27–35.

Jones, J.G. and Simon, B.M. (1981) Differences in microbial decomposition processes in profundal and littoral lake sediments, with particular reference to the nitrogen cycle. *Journal of General Microbiology*, **123**, 297–312.

Jones, J.G., Simon, B.M. and Horsley, R.W. (1982) Microbial sources of ammonia in freshwater lake sediments. *Journal of General Microbiology*, **128**, 2823–31.

Jones, J.R. and Bachman, R.W. (1976) Prediction of phosphorus and chlorophyll levels in lakes. *Journal of the Water Pollution Control Federation.*, **48**, 2176–82.

Jones, R., Benson-Evan, K., Chambers, F.M., Seddon, B.A. and Tai, Y.C. (1978) Biological and chemical studies of sediments from Llangorse Lake, Wales. *Verhandlungen Internationale Vereinigung für Theoretische und Angewandte Limnologie*, **20**, 642–8.

Jones, R.C., Walti, K. and Adams, M.S. (1983) Phytoplankton as a factor in the decline of the submerged macrophyte *Myriophyllum spicatum* L. in Lake Wingra, Wisconsin, USA *Hydrobiologia*, **107**, 213–9.

Jorgensen, S.E. (1976) A eutrophication model for a lake. *Ecological Modelling*, **2**, 147–65.

Joyce, J.C. (1985) Aquatic plant management – the Florida experience, in *Lake and Reservoir Management, Practical Applications*, North American Lake Management Society, McAfee, New Jersey, pp. 375–7.

Jupp, B.P. and Spence, D.H.N. (1977) Limitations on macrophytes in a eutrophic lake, Loch Leven. I. Effects of phytoplankton. *Journal of Ecology*, **65**, 175–86.

Kamp-Nielsen, L. (1985) Modelling of eutrophication processes, in *Lake Pollution and Recovery*, (eds. R. Vismara, R. Marforio, V. Mezzanotte, S. Cernuschi), European Water Pollution Control Association, Rome, pp. 57–69.

Kaushik, N.K., Robinson, J.B., Stammers, W.N. and Whiteley, H.R. (1981) Aspects of nitrogen transport and transformation in headwater streams, in *Perspectives in Running Water Ecology* (eds. M.A. Lock and D.D. Williams), Plenum Press, New York, pp. 113–40.

Kerekes, J.J. (1991) Aquatic birds as trophic indicators of water bodies.

Verhandlungen Internationale Vereinigung für Theoretische und Ange-wandte Limnologie, **24**. Kerfoot, W.C., DeMott, W.R. and DeAngelis, D.L. (1985) Interactions among cladocerans: food limitation and exploitative competition. *Archiv für Hydrobiologie Ergebnisse der Limnologie*, **21**, 431–51.

Kerr, P.C., Paris, D.F. and Brockway, D.L. (1970) *The interrelation of carbon and phosphorus in regulating heterotrophic and autotrophic populations in aquatic ecosystems.* US Government Printing Office, 16050 FGS.

Kerr, P.C., Brockway., D.L., Paris, D.F. and Barnett, J.T. (1972) The interrelation of carbon and phosphorus in regulation heterotrophic and autotrophic populations in an aquatic ecosystem, Shriner's Pond, in *Nutrients and Eutrophication: the limiting nutrient controversy* (ed. G.E. Likens), American Society for Limnology and Oceanography, Lawrence, Kansas, pp. 41–62.

Kerrison, P.H., McEwen, B., Phillips, G.L. and Crook, B.V. (1989) The use of redox potential to control ferric sulphate dosing during phosphate removal. *Journal of the Institution of Water Engineers and Scientists*, **43**, 397–403.

Kilham, P. and Hecky, R.E. (1988) Comparative Ecology of marine and freshwater phytoplankton. *Limnology and Oceanography*, **33**, 776–95.

King, D.L. (1970) The role of carbon in eutrophication. *Journal of the Water Pollution Control Federation*, **42**, 2035–51.

King, D.L. (1972) Carbon limitation in sewage lagoons, in *Nutrients and Eutrophication: the limiting nutrient controversy* (ed. G.E. Likens), American Society for Limnology and Oceanography, Lawrence, Kansas, pp. 98–110.

Kirchner, W.B. and Dillon, P.J. (1975) An empirical method of estimating the retention of phosphorus in lakes. *Water Resources Research*, **11**, 182–3.

Kloetzli, F. (1982) Some aspects of conservation in the overcultivated areas of the Swiss Midlands, in *Wetlands, Ecology and Management* (eds. B. Gopal, R.E. Turner, R.G. Wetzel and D.F. Whigham), International Scientific Publications, Jaipur, India, pp. 15–20.

Kloetzi, F. and Zust, S. (1973) Nitrogen regime in reed beds. *Polskie Archivum Hydrobiologia*, **20**, 131–6.

Knoechel, R. and Kalff, J. (1975) Algal sedimentation: the cause of a diatom – blue-green succession. *Verhandlungen Internationale Vereini-gung für Theoretische und Angewandte Limnologie*, **19**, 745–54.

Knoechel, R. and Kalff, J. (1978) An *in situ* study of the productivity and population dynamics of five freshwater planktonic diatom species. *Limnology and Oceanography*, **23**, 195–218.

Kohl, D.H., Shearer, G.B. and Commoner, B. (1971) Fertiliser nitrogen: contribution to nitrate in surface water in a corn belt watershed. *Science*, **174**, 1331–4.

Kohlenbrander, G.J. (1972) The eutrophication of surface water by agriculture and the urban population. *Stikstof*, **15**, 56–67.

Kratzer, C.R. and Brezonik, P.L. (1981) A Carlson-type trophic state index for nitrogen in Florida lakes. *Water Resources Bulletin*, **17**, 713–17.

Krebs, C.R. (1989) *Ecological Methodology*. Harper and Row Publishers, New York.

Kuenzel, L.E. (1969) Bacteria, carbon dioxide and algal blooms. *Journal of the Water Pollution Control Federation*, **41**, 1737–47.

Lachavanne, J. (1982) Influence de l'eutrophication des eaux sur les macrophytes des lacs Suisses: résultats préliminaires, in *Studies on Aquatic Vascular Plants* (eds. J.J. Symoens, S.S. Hooper and P. Compère), Royal Botanical Society of Brussels, Brussels, pp. 333–9.

Lambert, J.M., Jennings, J.N., Smith, C.T., Green, C. and Hutchinson, J.N. (1960) *The making of the Broads: a Reconstruction of their Origin in the Light of New Evidence*. Royal Geographical Society, London, Memoir No. 3.

Lambou, V.W., Taylor, W.D., Hern, S.C. and Williams, L.R. (1983) Comparison of trophic state measurements. *Water Research*, **17**, 1619–26.

Lammens, E.H.R.R. (1989) Causes and consequences of the success of bream in Dutch eutrophic lakes. *Hydrobiological Bulletin*, **23**, 11–8.

Lampert, W. and Muck, P. (1985) Multiple aspects of food limitation in zooplankton communities: the *Daphnia-Eudiaptomus* example. *Archiv für Hydrobiologie Ergebnisse der Limnologie*, **21**, 311–22.

Lampert, W. and Schober, U. (1980) The importance of 'threshold' food concentrations, in *Evolution and Ecology of Zooplankton Communities* (ed. W. Kerfoot), University Press of New England, Hanover, New Hampshire, pp. 264–7.

Landers, D.H. (1982) Effects of naturally senescing aquatic macrophytes on nutrient chemistry and chlorophyll a of surrounding waters. *Limnology and Oceanography*, **27**, 428–39.

Langford, P. (1979) Leisure and sports facilities at Rutland Water. *Journal of the Institution of Water Engineers and Scientists*, **39**, 117–34.

Larcher, W. (1975) *Physiological Plant Ecology*. Springer-Verlag, Berlin.

Larkin, P.A. and Northcote, T.G. (1969) Fish as indicators of eutrophication, in *Eutrophication: Causes, Consequences, Correctives* (ed. G.A. Rohlich), National Academy of Sciences, Washington DC, pp. 256–73.

Larsen, D.P. and Mercier, H.T. (1976) Phosphorus retention capacity of lakes. *Journal of the Fisheries Research Board of Canada*, **33**, 1742–50.

Lasenby, D.C. (1975) Development of oxygen deficits in 14 Southern Ontario lakes. *Limnology and Oceanography*, **20**, 993–9.

Laws, E.A. (1981) *Aquatic Pollution*. Wiley-Interscience, New York.

Leach, J.H., Johnson, M.G., Kelso, J.R.M., Hartmann, J., Nümann, W. and Entz, B. (1977) Responses of percid fishes and their habitats to eutrophication. *Journal of the Fisheries Research Board of Canada*, **34**, 1964–71.

Leah, R.T., Moss, B. and Forrest, D.E. (1978) Experiments with large enclosures in a fertile, shallow, brackish lake, Hickling Broad, Norfolk, United Kingdom. *Internationale Revue der Gesampten Hydrobiologie*, **63**, 291–310.

Lean, D.R.S. (1973a) Phosphorus dynamics in lake water. *Science*, **179**, 678–80.

Lean, D.R.S. (1973b) Movements of phosphorus between its biologically important forms in lake water. *Journal of the Fisheries Research Board of Canada*, **30**, 1525–36.

Lean, D.R.S. and Nalewajko, C. (1975) Phosphate exchange and organic phosphorus excretion by algae. *Journal of the Fisheries Research Board of Canada*, **32**, 1312–23.

Lee, G.F. and Jones, R.A. (1986) Detergent phosphate bans and eutrophication. *Environmental Science and Technology*, **20**, 330–1.

Lehman, J.T. (1986) The goal of understanding in limnology. *Limnology and Oceanography*, **31**, 1160–1166.

Lehman, J.T. and Sandgren, C.D. (1985) Species-specific rates of growth and grazing loss among freshwater algae. *Limnology and Oceanography*, **30**, 34–46.

Leopold, M., Bninínska, M. and Nowak, W. (1986) Commercial fish catches as an index of lake eutrophication. *Archiv für Hydrobiologie*, **106**, 513–24.

Levine, S.N. and Schindler, D.W. (1989) Phosphorus, nitrogen and carbon dynamics of experimental lake 303 during recovery from eutrophication. *Canadian Journal of Fisheries and Aquatic Sciences*, **46**, 2–10.

Lewis, W.M.J. (1974) Primary production in the plankton community of a tropical lake. *Ecological Monographs*, **44**, 377–409.

Likens, G.E. (1972) *Nutrients and Eutrophication: the Limiting Nutrient Controversy*. American Society of Limnology and Oceanography, Lawrence, Kansas.

Likens, G.E. (1984) Beyond the shoreline: a watershed ecosystem approach. *Verhandlungen Internationale Vereinigung für Theoretische und Angewandte Limnologie*, **22**, 1–22.

Lindeman, R.L. (1942) The tropho-dynamic aspect of ecology. *Ecology*, **23**, 399–418.

Livingstone, D.A. (1984) The preservation of algal remains in lake sediments, in *Lake sediments and environmental history* (eds. E.Y. Haworth and J.W.G. Lund), Leicester University Press, Leicester, pp. 191–202.

Löfgren, S. and Ryding, S. (1985) Apatite solubility and microbial activities as regulators of internal loading in shallow, eutrophic lakes. *Verhandlungen Internationale Vereinigung für Theoretische und Angewandte Limnologie*, **22**, 3329–34.

Lopez, G.R. (1988) Comparative ecology of the macrofauna of freshwater and marine muds. *Limnology and Ocaenography*, **33**, 946–2.

Lorensen, M.W. and Fast, A.W. (1977) *A guide to aeration/circulation techniques for lake management*. U.S. Environmental Protection Agency,

Washington DC, Research Series EPA 600/3–77–004.

Lund, J.W.G. (1950) Studies on *Asterionella formosa* Hass. II. Nutrient depletion and the spring maximum. *Journal of Ecology*, **38**, 1–35.

Lund, J.W.G. (1969) Phytoplankton, in *Eutrophication: Causes, Consequences, Correctives* (ed. G.A. Rohlich), National Academy of Sciences, Washington DC, pp. 306–30.

Lund, J.W.G. (1970) Primary Production. *Water Treatment and Examination*, **19**, 332–58.

Lund, J.W.G. (1972) Eutrophication. *Proceedings of the Royal Society of London*, B, **180**, 371–82.

Lund, J.W.G., Kipling, C. and Le Cren, E.D. (1963) Changes in depth and time of certain chemical and physical conditions and of the standing crop of *Asterionella formosa* Hass. in the North Basin of Windermere in 1947. *Philosophical Transactions of the Royal Society of London*, B, **246**, 255–90.

Lynch, M. (1977) Fitness and optimal body size in zooplankton populations. *Ecology*, **58**, 763–74.

Lynch, M. and Shapiro, J. (1981) Predation, enrichment and phytoplankton community structure. *Limnology and Oceanography*, **26**, 86–102.

Macan, T.T. (1970) *Biological Studies of the English Lakes*. Longman Group, London.

Macan, T.T. (1980) Changes in the fauna of the stony substratum of lakes in the English Lake District. *Hydrobiologia*, **72**, 159–64.

MacArthur, R.H. and Wilson, E.O. (1967) *The Theory of Island Biogeography*. Princeton University Press, Princeton, New Jersey.

Mackereth, F.J.H. (1966) Some chemical observations on post-glacial lake sediments. *Philosophical Transactions of the Royal Society of London*, B, **250**, 165–213.

Magarara, Y. and Kunikane, S. (1986) Cost analysis of the adverse effects of algal growth in water bodies on drinking water supply. *Ecological Modelling*, **31**, 303–13.

Makarewicz, J.C. and Likens, G.E. (1979) Structure and function of the zooplankton community of Mirror Lake, New Hampshire. *Ecological Monographs*, **49**, 109–27.

Maki, A.W. (1984) The impact of detergent phosphorus bans on receiving water quality. *Water Resources*, **18**, 893–903.

Maloney, T.E., Miller, W.E. and Shiroyama, T. (1972) Algal responses to nutrient additions in natural waters. I. Laboratory bioassays, in *Nutrients and eutrophication: the limiting nutrient controversy* (ed. G.E. Likens), American Society of Limnology and Oceanography, Lawrence, Kansas, pp. 134–56.

Malthus, T.J. and Mitchell, S.F. (1988) Agricultural development and eutrophication of Lake Mahinerangi, New Zealand. *Verhandlungen Internationale Vereinigung für Theoretische und Angewandte Limnologie*, **23**, 1028–31.

Mamamah, D.S. and Bhagat, S.K. (1982) Performance of some empirical phosphorus models. *Proceedings of the American Society of Civil Engineers, Journal of the Environmental Engineering Division*, **108**, 722–729.

Mann, K.H. (1980) Benthic secondary production. In *Fundamentals of Aquatic Ecosystems* (eds. R.S.K. Barnes and K.H. Mann), Blackwell Scientific Publishers, Oxford, pp. 103–18.

Manny, B.A. (1972) Seasonal changes in dissolved organic nitrogen in six Michigan lakes. *Verhandlungen Internationale Vereinigung für Theoretische und Angewandte Limnologie*, **18**, 147–56.

Martin, W.P., Fenster, W.E. and Hanson, L.D. (1970) Fertiliser management for pollution control, in *Agriculture and Water Quality* (eds. T. Willrich and G.E. Smith), Iowa University Press, Iowa, pp. 142–58.

Marsden M.W. (1989) Lake restoration by reducing external phosphorus loading: the influence of sediment phosphorus release. *Freshwater Biology*, **21**, 139–62.

Mason, C.F. (1976) Broadland, in *Nature in Norfolk: a Heritage in Trust* (ed. R. Washbourn), Jarrold, Norwich, pp. 79–89.

Mason, C.F. (1977) Populations and production of benthic animals in two contrasting shallow lakes in Norfolk. *Journal of Animal Ecology*, **46**, 147–72.

Mason, C.F. (1981) *Biology of Freshwater Pollution*. Longmans, London.

Mason, C.F. and Bryant, R.J. (1975) Changes in the ecology of the Norfolk Broads. *Freshwater Biology*, **5**, 257–70.

McCarty, P.L. and Beck, L. (1969) Biological denitrification of waste waters by addition of organic materials. *Proceedings of the 24th Institute of Wastes Conference*, Purdue University, Purdue, pp. 1271–85.

McCauley, E. and Briand, F. (1979) Zooplankton grazing and phytoplankton species richness: field tests of the predation hypothesis. *Limnology and Oceanography*, **24**, 243–52.

McCauley, E. and Kalff, J. (1981) Empirical relationships between phytoplankton and zooplankton in lakes. *Canadian Journal of Fisheries and Aquatic Science*, **38**, 458–63.

McErlean, A.J. and Reed, G. (1981) Indicators and indices of estuarine overenrichment, in *Estuaries and Nutrients* (eds. B.J. Neilson and L.E. Cronin), Humana Press, Clifton, New Jersey, pp. 165–82.

McNaught, D.C. (1975) A hypothesis to explain the succession form calanoids to cladocerans during eutrophication. *Verhandlungen Internationale Vereinigung für Theoretische und Angewandte Limnologie*, **19**, 17–31.

McQueen, D.J. (1990) Manipulating lake community structure: where do we go from here? *Freshwater Biology*, **23**, 613–20.

McQueen, D.J., Post, J.R. and Mills, E.L. (1986) Trophic relationships in freshwater pelagic ecosystems. *Canadian Journal of Fisheries and Aquatic Sciences*, **43**, 1571–81.

Meijer, M.L., Raat, A.J.P. and Doef, R.W. (1989) Restoration by biomanipulation of Lake Bleiswijke Zoom (The Netherlands): First results. *Hydrobiological Bulletin,* **23,** 49–57.

Mekerson, K.A. (1973) Phosphorus in chemical and physical treatment processes. *Water Research,* **7,** 145–58.

Meybeck, M. (1982) Carbon, Nitrogen and Phosphorus transport by world rivers. *American Journal of Science,* **282,** 401–50.

Miller, W.E., Greene, J.C. and Shiroyama, T. (1978) *The* Selenastrum capricornutum *Printz Algal Assay Bottle Test.* United States Environmental Protection Agency, Environmental Research Laboratory, Corvallis, Oregon, Report EPA 600/9-78-018.

Montgomery, J.M.C.E., Inc. (1985) *Water Treatment Principles and Design.* John Wiley and Sons, New York.

Moore, D.E. (1982) Establishing and maintaining the trout fishery at Rutland Water. *Hydrobiologia,* **88,** 179–89.

Moore, N.W. (1987) *The Bird of Time – The Science and Politics of Nature Conservation.* Cambridge University Press, Cambridge.

Morgan, N.C. (1970a) Problems of the conservation of freshwater ecosystems, in *Conservation and Productivity of Natural Waters* (eds. R.W. Edwards and D.J. Garrod), Academic Press, London, pp. 115–33.

Morgan, N.C. (1970b) Changes in the flora and fauna of a nutrient enriched lake. *Hydrobiologia,* **35,** 545–53.

Morgan, N.C. and McLusky, D.S. (1974) A summary of the Loch Leven IBP results in relation to lake management and future research. *Proceedings of the Royal Society of Edinburgh,* B., **74,** 407–16.

Mortimer, C.H. (1941) The exchange of dissolved substances between mud and water in lakes. *Journal of Ecology,* **29,** 280–329.

Mortimer, C.H. (1942) The exchange of dissolved substances between mud and water in lakes. *Journal of Ecology,* **30,** 147–201.

Moss, B. (1969) Limitation of algal growth in some Central African waters. *Limnology and Oceanography,* **14,** 591–601.

Moss, B. (1972) Studies on Gull Lake Michigan. II Eutrophication – evidence and prognosis. *Freshwater Biology,* **2,** 309–20.

Moss, B. (1977) Conservation problems in the Norfolk Broads and rivers of East Anglia – phytoplankton, boats and causes of turbidity. *Biological Conservation,* **10,** 261–79.

Moss, B. (1979a) Limitation of algal growth in some Central African waters. *Limnology and Oceanography,* **14,** 591–601.

Moss, B. (1979b) Algal and other fossil evidence for major changes in Strumpshaw Broad, Norfolk, England in the last two centuries. *British Phycological Journal,* **14,** 263–83.

Moss, B. (1980) Further studies on the paleolimnology and changes in the phosphorus budget of Barton Broad, Norfolk. *Freshwater Biology,* **10,** 261–79.

Moss, B. (1988) *Ecology of Freshwaters: Man and Medium* (2nd ed.) Blackwell Scientific Publishers, Oxford.

Moss, B. (1989) Water pollution and the management of ecosystems; a case study of science and scientist, in *Toward a More Exact Ecology* (eds. P.J. Grubb and J.B. Whittaker), Blackwell Scientific Publications, Oxford, pp. 401–22.

Moss, B., Forrest, D. and Phillips, G.L. (1979a) Eutrophication and paleolimnology of two small medieval man-made lakes. *Archiv für Hydrobiologie*, 85, 409–25.

Moss, B., Leah, R.T. and Clough, B. (1979b) Problems of the Norfolk Broads and their impact on Freshwater Fisheries, in *Proceedings of the First Freshwater Fisheries Conference* (eds. K. O'Hara and C. Dickson), University of Liverpool, pp. 67–85.

Moss, B., Balls, H., Irvine, K. and Stansfield, J. (1986) Restoration of two lowland lakes by isolation from nutrient-rich water sources with and without removal of sediment. *Journal of Applied Ecology*, 23, 391–414.

Moss, B., Balls, H., Booker, I., Manson, K. and Timms, M. (1988) Problems in the construction of a nutrient budget for the River Bure and its Broads (Norfolk) prior to its restoration from eutrophication, in *Algae and the Aquatic Environment* (ed. F.E. Round), Biopress, Bristol, pp. 326–53.

Mouchet, P.C. (1984) Influence of recently drowned terrestrial vegetation on the quality of water stored in impounding reservoirs. *Verhandlungen Internationale Vereinigung für Theoretische und Angewandte Limnologie*, 22, 1608–19.

Mueller, D.K. (1982) Mass balance model estimations of phosphorus concentrations in reservoirs. *Water Resources Bulletin*, 18, 377–82.

Müller, D. and Kirchesch, V. (1982) Hypertrophy in slow flowing rivers. *Developments in Hydrobiology*, 2, 338.

Murray, D.A. (1979) The evolution of pollution evidenced by lake sediment pseudofossils, in *Biological Aspects of Freshwater Pollution* (ed. O. Ravera), Pergamon Press, Oxford, pp. 77–91.

Naumann, E. (1919) Nagra synpunkte angaende planktons okologi. Med. sarskilde hansyn till fytoplankton. *Svensk Botaniske Tidskrift*, 13, 129–58.

Negoro, K. (1981) Studies on the phytoplankton of Lake Biwa. *Verhandlungen Internationale Vereinigung für Theoretische und Angewandte Limnologie*, 21, 574–83.

Neill, W.E. and Peacock, A. (1980) Breaking the bottleneck: interactions of invertebrate predators and nutrients in oligotrophic lakes, in *Evolution and Ecology of Zooplankton Communities* (ed. W.C. Kerfoot), University Press of New England, Hanover, New Hampshire, pp. 715–24.

Newbold, C., Fry, G. and Cooke, A.S. (1986) Tackling agricultural pollution – the difficulties facing conservationists in Great Britain, in *Effects of Land Use on Freshwaters: Agriculture, Forestry, Mineral Exploitation, Urbanisation* (ed. J.F. de.L.G. Solbé), Ellis Horwood Ltd., Chichester, pp. 296–314.

Nicholls, K.H. and Dillon, P.J. (1978) An evaluation of phosphorus-chlorophyll-phytoplankton relationships for lakes. *Internationale Revue der Gesampten Hydrobiologie*, **63**, 141–54.

Nicholson, W.A. (1900) Sir Thomas Browne as a Naturalist. *Transactions of the Norfolk and Norwich Naturalists' Society*, **7**, 72–89.

de Nie, H.W. (1987) *The decrease in aquatic vegetation in Europe and its consequences for fish populations*. European Inland Fisheries Advisory Commission, EIFAC/CECPI. No. 19.

Nilssen, J.P. (1978) Eutrophication, minute algae and inefficient grazers. *Memorie dell'Instituto Italiano di Idrobiologia*, **36**, 121–38.

Nümann, W. (1973) Versuch einer Begründung für den Wandel in der qualitativen und quantitativen Zusammensetzung des Fischbestandes im Bodensee während der letzen 60 jahre und eine Bewertung der Besatzmassnahmen. *Schweizerische Zeitschrift fuer Hydrologie*, **35**, 206–38.

Nye, P.H. and Greenland, D.J. (1960) *The Soil Under Shifting Cultivation*. Commonwealth Agricultural Bureaux. Farnam Royal, Buckinghamshire, England.

Nygaard, G. (1949) Hydrobiological studies of some Danish lakes and ponds. *Biologiske Skrifter*, **7**, 1–293.

O'Brien, W.J. (1972) Limiting factors in phytoplankton algae: their meaning and measurement. *Science*, **178**, 616–7.

O'Neill, J.G. and Holding, A.J. (1975) The importance of nitrate reducing bacteria in lakes and reservoirs, in *The Effects of Storage on Water Quality*, Water Research Centre, Medmenham, Marlow, pp. 91–123.

Odum, E.P. (1971) *Fundamentals of Ecology* (3rd ed.). W.B. Saunders, Philadelphia.

Oglesby, R.T. (1969) Effects of controlled nutrient dilution on the eutrophication of a lake, in *Eutrophication: Causes, Consequences, Correctives*, (ed. G.A. Rohlich), National Academy of Sciences, Washington, DC, pp. 483–93.

Oglesby, R.T. (1977) Relationships of fish yield to lake phytoplankton standing crop, production and morphoedaphic factors. *Journal of the Fisheries Research Board of Canada*, **34**, 2271–9.

Oldfield, F. (1977) Lakes and their drainage basins as units of sediment-based ecological study. *Progress in Physical Geography*, **1**, 460–504.

Orlob, G.T. (1988) Mathematical models of lakes and reservoirs, in *Ecosystems of the World: Lakes and Reservoirs* (ed. F.B. Taub), Elsevier, Amsterdam, pp. 43–62.

Osborne, P.L. (1978) *Relationships between the Phytoplankton and Nutrients in the River Ant and Barton, Sutton and Stalham Broads, Norfolk*. Ph.D. thesis, University of East Anglia.

Osborne, P.L. (1980) Prediction of phosphorus and nitrogen concentrations in lakes from both internal and external loading rates. *Hydrobiologia*, **69**, 229–33.

Osborne, P.L. (1981) Phosphorus and nitrogen budgets of Barton Broad and predicted effects of a reduction in nutrient loading on phytoplankton biomass in Barton, Sutton and Stalham Broads, Norfolk, United Kingdom. *Internationale Revue der Gesampten Hydrobiologie*, **66**, 171–202.

Osborne, P.L. and Moss, B. (1977) Paleolimnology and trends in the phosphorus and iron budgets of an old man-made lake. *Freshwater Biology*, **7**, 213–33.

den Oude, P.J. and Gulati, R.G. (1988) Phosphorus and nitrogen excretion rates of zooplankton from the eutrophic Loosdrecht lakes, with notes on other P sources for phytoplankton requirements. *Hydrobiologia*, **169**, 379–90.

Owens, M. (1970) Nutrient balances in rivers. *Water Treatment and Examination*, **19**, 239–52.

Owens, M. and Wood, G. (1968) Some aspects of the eutrophication of water. *Water Research*, **2**, 151–9.

Pace, M.L. (1986) An empirical analysis of zooplankton community size structure across lake trophic gradients. *Limnology and Oceanography*, **31**, 45–55.

Paerl, H.W. (1988) Nuisance phytoplankton blooms in coastal, estuarine and inland waters. *Limnology and Oceanography*, **33**, 823–47.

Paerl, H.W., Tucker, J. and Bland, P.T. (1983) Carotenoid enhancement and its role in maintaining blue-green algal (*Microcystis aeruginosa*) surface blooms. *Limnology and Oceanography*, **28**, 847–57.

Pallesen, L., Berthouex, P.M. and Booman, K. (1985) Environmental intervention analysis: Wisconsin's ban on phosphate detergents. *Water Research*, **19**, 353–62.

Pallis, M. (1911) The river valleys of East Norfolk: their aquatic and fen formations, in *Types of British Vegetation* (ed. A.G. Tansley), Cambridge University Press, Cambridge, pp. 214–45.

Palmer, C.M. (1969) A composite rating of algae tolerating organic pollution. *Journal of Phycology*, **5**, 78–82.

Paloheimo, J.E. and Zimmerman, A.P. (1983) Factors influencing phosphorus-phytoplankton relationships. *Canadian Journal of Fisheries and Aquatic Sciences*, **40**, 1804–12.

Park, R.A. and O'Neill, R.V. (1974) A generalised model for simulating lake ecosystems. *Simulation*, **21**, 33–50.

Parker, D.J. and Penning-Rowsell, E.C. (1980) *Water Planning in Britain*. George Allen and Unwin, London.

Pastorak, R.A. (1980) Selection of prey by Chaoborus larvae: a review and new evidence for behavioural flexibility, in *Evolution and Ecology of Zooplankton Communities* (ed. W.C. Kerfoot, University Press of New England, Hanover, New Hampshire, pp. 538–54.

Pastorak, R.A., Ginn, T.C. and Lorenzen, M.W. (1980) Review of aeration/circulation for lake management, in *Restoration of Lakes and Inland Waters*. U.S. Environmental Protection Agency, Washington, DC, EPA 440/5–81–010, pp. 124–33.

Patalas, K. (1972) Crustacean plankton and the eutrophication of the St. Lawrence Great Lakes. *Journal of the Fisheries Research Board of Canada*, **29**, 1451–62.

Payne, A.E. (1984) Use of sewage waste in warm water aquaculture. *Institute of Civil Engineers Symposium on the Re-use of Sewage Effluent*, Thomas Telford Press, London, pp. 117–31.

Pearsall, W.H. (1921) The development of vegetation in the English lakes, considered in relation to the general evolution in glacial lakes and rock basins. *Proceedings of the Royal Society of London*, Series B, **92**, 259–84.

Pearsall, W.H. (1924) Phytoplankton and environment in the English Lake District. *Revue Algologique*, **1**, 53–67.

Pearsall, W.H. (1930) Phytoplankton in the English Lakes 1. The proportions in the water of some dissolved substances of biological importance. *Journal of Ecology*, **18**, 306–20.

Pearsall, W.H. (1932) Phytoplankton in the English Lakes. II. The composition of the phytoplankton in relation to dissolved substances. *Journal of Ecology*, **30**, 241–62.

Pearsall, W.H., Gardiner, A.C. and Greenshields, F. (1946) Freshwater Biology and Water Supply in Britain. *Scientific Publications of the Freshwater Biological Association*, **11**, FBA, Ambleside, Cumbria.

Pennington, W. (1981) Records of a lake's life in time: the sediments. *Hydrobiologia*, **79**, 197–219.

Persson, P.E. (1980) Muddy odour in fish from hypertrophic waters. *Developments in Hydrobiology*, **2**, 203–8.

Persson, G. and Jansson, M. (1988) *Phosphorus in Freshwater Ecosystems*. Kluwer Academic Publishers, Dortrecht, Holland.

Peterjohn, W.T. and Correll, D.L. (1984) Nutrient dynamics in an agricultural watershed: observations on the role of a riparian forest. *Ecology*, **65**, 1466–75.

Peters, R.H. (1975) Phosphorus regeneration by natural populations of limnetic zooplankton. *Verhandlungen Internationale Vereinigung für Theoretische und Angewandte Limnologie*, **19**, 273–9.

Peters, R.H. (1984) Methods for the study of feeding, grazing and assimilation by zooplankton, in *A manual on Methods for the Assessment of Secondary Productivity in Freshwaters* (eds. J.A. Downing and F.H. Rigler), Blackwell Scientific Publications, Oxford, pp. 336–412.

Peters, R.H. (1986) The role of prediction in limnology. *Limnology and Oceanography*, **31**, 1143–59.

Peters, R.H. and Downing, J.A. (1984) Empirical analysis of zooplankton filtering and feeding rates. *Limnology and Oceanography*, **29**, 763–84.

Peterson, S.A. (1979) Dredging and lake restoration, in *Lake Restoration: Proceedings of a national conference*, US Environmental Protection Agency, Washington, DC, pp. 105–14.

Phillips, S.P. (1963) A note on the Charophytes of Hickling Broad, E. Norfolk. *Proceedings of the Botanical Society of the British Isles*, **5**, 23–4.

Phillips, G.L. (1984) A large scale field experiment in the control of eutrophication in the Norfolk Broads. *Journal of the Institute of Water*

Pollution Control, **83**, 400–8.

Phillips, G.L. and Jackson, R. (1989) *The Importance of Sediment Release of Phosphorus in the Restoration of the Norfolk Broads*, Nature Conservancy Council, Norwich, Unpublished report, Contract Number HF3–03–350.

Phillips, G.L. and Jackson, R. (in press) The control of eutrophication in very shallow lakes, the Norfolk Broads. *Verhandlungen Internationale Vereinigung Theoretische und Angewandte Limnologie*, **24**.

Phillips, G.L. and Moss, B. (1978) *The Distribution, Biomass and Productivity of Submerged Aquatic Plants in the Thurne Broads, Norfolk*, Nature Conservancy Council, Norwich, unpublished report.

Phillips, G.L., Eminson, D.F. and Moss, B. (1978) A mechanism to account for macrophyte decline in progressively eutrophicated freshwaters. *Aquatic Botany*, **4**, 103–26.

Pierrou, U. (1976) The global phosphorus cycle, in *Nitrogen, Phosphorus and Sulphur – Global Cycles* (eds. B.H. Svensson and R. Söderlund), Ecological Bulletins 22, Stockholm, pp. 75–88.

Pitcher, T.J. and Hart, P.J.B. (1985) *Fisheries Ecology*. Croom Helm, London.

Porcella, D.B. (1985) The effects of changes in external phosphorus loadings, in *Lake pollution and Recovery* (eds. R. Vismara, R. Marforio, V. Mezzanotte and S. Cernuschi). Associazione Nationale di Ingegneria Sanitaria, Rome, pp. 125–32.

Porcella, D.B. and Bishop, A.B. (1976) *Comprehensive Management of Phosphorus Pollution*. Ann Arbor Science, Ann Arbor, Michigan.

Porter, K.G. (1977) The plant-animal interface in freshwater ecosystems. *American Scientist*, **65**, 159–70.

Porter, K.G., Gerritzen, J. and Orcutt, J.D.J. (1982) The effect of food concentration on swimming patterns, feeding behaviour, ingestion, assimilation and respiration by *Daphnia*. *Limnology and Oceanography*, **27**, 935–49.

Porter, K.G. and Orcutt, J.R. (1980) Nutritional adequacy, manageability and toxicity as factors that determine the food quality of green and blue gree algae for *Daphnia*, in (ed. W.C. Kerfoot), University Press of New England, Hanover, New Hampshire, pp. 268–81.

Porter, K.S. (1975) *Nitrogen and Phosphorus; food production, waste and the environment*, Ann Arbor Science, Ann Arbor, Michigan.

Porter, K.S., Jacobs, J., Lauer, D.A. and Young, R.J. (1975a) Nitrogen and phosphorus in the environment, in *Nitrogen and Phosphorus; food production, waste and the environment* (ed. K.S. Porter), Ann Arbor Science, Ann Arbor, Michigan, pp. 3–21.

Porter, K.S., Lauer, D.A., Messinger, J. and Bouldin, D.R. (1975b) Flows of nitrogen and phosphorus on land, in *Nitrogen and Phosphorus; food production, waste and the environment* (ed. K.S. Porter), Ann Arbor Science, Ann Arbor, Michigan, pp. 123–65.

Pullianen, E. (1981) The history and spread of the moorhen, *Gallinula chloropus* in Finland. *Ornis Fennica*, **57**, 117–23.

Rasmussen, J.B. (1984) The life history, distribution, and production of *Chironomus riparius* and *Glyptotendipes paripes* in a prairie pond. *Hydrobiologia*, **119**, 65–72.

Ravera, O., Garavaglia, C. and Stella, M. (1984) The importance of the macrophytes in two lakes with different trophic degree: Lake Comabbio and Lake Monate (Province of Varese, Northern Italy). *Verhandlungen Internationale Vereinigung für Theoretische und Angewandte Limnologie*, **19**, 1119–30.

Raymont, J.E.G. (1980) *Plankton and Productivity in the Oceans* (2nd ed.), Pergamon Press, Oxford.

Reckhow, K.H. (1979a) Uncertainty applied to Vollenweider's phosphorus criterion. *Journal of the Water Pollution Control Federation*, **51**, 2123–8.

Reckhow, K.H. (1979b) Empirical lake models for phosphorus: development, applications, uncertainty, in *Perspectives on Lake Ecosystem Modeling* (eds. D. Scavia and A. Robertson), Ann Arbor Science, Ann Arbor, Michigan, pp. 193–222.

Reckhow, K.H. and Simpson, J.T. (1980) A procedure using modeling and error analysis for the prediction of lake phosphorus concentration from land use information. *Canadian Journal of Fisheries and Aquatic Sciences*, **37**, 1439–48.

Reckhow, K.H., Beaulac, M.N. and Simpson, J.T. (1980) *Modelling phosphorus loading in lake response under uncertainty: a manual and compilation of export coefficients*. US Environmental Protection Agency, Washington, DC, EPA 440/5–80–011.

Redfield, A.C. (1934) On the proportions of organic derivatives in sea water and their relation to the composition of plankton. *James Johnstone Memorial Volume*, Liverpool University Press, Liverpool, pp. 176–92.

Regier, H.A. and Hartmann, W.L. (1973) Lake Erie's fish community: 150 years of cultural stress. *Science*, **180**, 1248–55.

Reynolds, C.S. (1979) The limnology of the eutrophic meres of the Shropshire-Cheshire plain: a review. *Field Studies*, **5**, 93–173.

Reynolds, C.S. (1980) Phytoplankton assemblages and their periodicity in stratifying lake systems. *Holarctic Ecology*, **3**, 141–59.

Reynolds, C.S. (1982) Phytoplankton periodicity: its motivation, mechanisms and manipulation. *Annual Report of the Freshwater Biological Association*, **50**, 60–75.

Reynolds, C.S. (1984a) *The Ecology of Freshwater Phytoplankton*. Cambridge University Press, Cambridge.

Reynolds, C.S. (1984b) Phytoplankton periodicity: the interactions of form, function and environmental variability, *Freshwater Biology*, **14**, 111–42.

Reynolds, C.S. and Wiseman, S.W. (1982) Sinking losses of phytoplankton in closed limnetic systems. *Journal of Plankton Research*, **4**, 489–522.

Reynolds, C.S., Wiseman, S.W. and Clarke, M.J.O. (1984) Growth- and loss-rate responses of phytoplankton to intermittent mixing and their potential application to the control of planktonic algal biomass. *Journal of Applied Ecology*, **21**, 11–39.

Reynolds, C.S., Wiseman, S.W. and Gardiner, W.D. (1980) *Aquatic Sediment Traps and Trapping Methodology*. Freshwater Biological Association Occasional Publications 11, Windermere.

Reynolds, C.S., Thompson, J.M., Ferguson, A.J.D. and Wiseman, S.W. (1982) Loss processes in the population dynamics of phytoplankton maintained in closed systems. *Journal of Plankton Research*, **4**, 561–600.

Reynolds, C.S., Wiseman, S.W., Godfrey, B.M. and Butterwick, C. (1983) Some effects of artificial mixing on the dynamics of phytoplankton populations in large limnetic enclosures. *Journal of Plankton Research*, **5**, 203–34.

Reynoldson, T.B.J. and Hamilton, H.R. (1982) Spatial heterogeneity in whole lake sediments – towards a loading estimate, *Hydrobiologia*, **91**, 235–40.

Rhee, G. (1978) Effects of N:P atomic ratios and nitrate limitation on algal growth, cell composition and nitrate uptake, *Limnology and Oceanography*, **23**, 10–25.

Rich, P.H., Wetzel, R.G. and Van Thuy, N. (1971) Distribution, production and role of aquatic macrophytes in a southern Michigan marl lake. *Freshwater Biology*, **1**, 3–21.

Richardson, J.L. and Richardson, A.E. (1972) History of an African Rift lake and its climatic implications. *Ecological Monographs*, **42**, 499–534.

Richardson, K., Beardall, J. and Raven, J.A. (1983) Adaptation of unicellular algae to irradiance: an analysis of strategies. *New Phytologist*, **93**, 157–1.

Richman, S. and Dodson, S.I. (1983) The effect of food quality of feeding and respiration by *Daphnia* and *Diaptomus*. *Limnology and Oceanography*, **28**, 948–56.

Ridley, J.E. (1975) Problems of emergent insect species associated with water storage, in *The Effects of Storage on Water Quality*, Water Research Centre, Medmenham, Marlow, pp. 273–81.

Riemann, B., Søndergaard, M., Persson, L. and Johansson, L. (1986) Carbon metabolism and community regulation in eutrophic, temperate lakes, in *Carbon Dynamics in Eutrophic, Temperate Lakes* (eds. B. Riemann and M. Søndergaard), Elsevier Science Publishers, Amsterdam, pp. 267–80.

Rigler, F.H. (1974) Phosphorus cycling in lakes, in *Fundamentals of Limnology* (ed. F. Ruttner), University of Toronto Press, Toronto, pp. 263–73.

Riley, G.A. (1946) Factors controlling phytoplankton populations on Georges Bank. *Journal of Marine Research*, **6**, 54–73.

Riley, G.A. (1947) A theoretical analysis of the zooplankton population of

Georges Bank. *Journal of Marine Research*, **6**, 104–13.

Ritter, W.R. (1985) Comparison of eutrophication indexes for Delaware lakes. *Transactions of the American Society of Agricultural Engineers*, **28**, 1591–7.

Robarts, R.D. and Zohary, T. (1984) *Microcystis aeruginosa* and underwater light attenuation in a hypereutrophic lake (Hartbeesport dam, South Africa) *Journal of Ecology*, **72**, 1001–7.

Rodhe, W. (1958) Primar produktion und seentypen. *Verhandlungen Internationale Vereinigung für Theoretische und Angewandte Limnologie*, **13**, 121–41.

Rodhe, W. (1965) Standard correlations between pelagic photosynthesis and light. *Memorie dell' Istituto Italaliano Idrobiologia* (Suppl.), **18**, 365–81.

Rodhe, W. (1969) Crystallisation of eutrophication concepts in northern Europe, in *Eutrophication – Causes, Consequences, Correctives* (ed. G.A. Rohlich), National Academy of Sciences, Washington, DC, pp. 65–97.

Rohlich, G.A. (1969) *Eutrophication – Causes, Consequences, Correctives*. National Academy of Sciences, Washington, DC.

Romanovsky, Y.E. (1985) Food limitation and life-history strategies in cladoceran crustaceans. *Archiv für Hydrobiologie Beiheft. Ergebnisse der Limnologie*, **21**, 363–72.

Room, P.R. (1990) Ecology of a simple plant-herbivore system: biological control of Savinia. *Trends in Ecology and Evolution*, **5**, 74–8.

Ross, P.E. and Kalff, J. (1975) Phytoplankton production in Lake Memphremagog, Québec (Canada) – Vermont (USA) *Verhandlungen Internationale Vereinigung für Theoretische und Angewandte Limnologie*, **19**, 760–9.

Round, F.E. (1981) *The Ecology of Algae*. Cambridge University Press, Cambridge.

Russell, E.J. and Richards, E.H. (1919) The amount and composition of rain falling at Rothamsted. *Journal of Agricultural Science, Cambridge*, **9**, 309–37.

Ryding, S. and Fosberg, C. (1982) Short-term load-response relationships in shallow, polluted lakes. *Hydrobiologia*, **87**, 95–103.

Ryding, S. and Rast, W. (1989) *The Control of Eutrophication of Lakes and Reservoirs*. UNESCO and the Parthenon Publishing Group, Paris.

Ryther, J.H. (1981) Impact of nutrient enrichment on water uses, in *Estuaries and Nutrients* (eds. B.J. Neilson and L.E. Cronin), Humana Press, Clifton, New Jersey, pp. 247–61.

Saad, M.A.H. (1980) Eutrophication of Lake Mariut, a heavily polluted lake in Egypt. *International Atomic Energy Agency Vienna*, 153–63.

Sakamoto, M. (1966) Primary production by the phytoplankton community in some Japanese lakes and its dependence upon lake depth. *Archiv für Hydrobiologie*, **62**, 1–28.

Sand-Jensen, K. and Borum, T. (1984) Epiphyte shading and its effect on photosynthesis and diel metabolism of *Lobelia dortmanna* L. during the spring bloom in a Danish lake. *Aquatic Botany*, **20**, 15–32.

Sand-Jensen, K. and Søndergard, M. (1981) Phytoplankton and epiphyte development and their shading effect on submerged macrophytes in lakes of different nutrient status. *Internationale Revue der Gesampten Hydrobiologie*, **66**, 529–52.

Sas, H. (1989) *Lake Restoration by Reduction of Nutrient Loading*. Academia Verlag Richarz GmbH, St. Augustin.

Sawyer, C.N. (1947) Fertilisation of lakes by agricultural and industrial drainage. *New England Water Works Association*, **61**, 109–27.

Sæther, O.A. (1975) Nearctic chironomids as indicators of lake typology. *Verhandlungen Internationale Vereinigung für Theoretische und Angewandte Limnologie*, **19**, 3127–33.

Sæther, O.A. (1979) Chironomid communities as water quality indicators. *Holarctic Ecology*, **2**, 65–74.

Sæther, O.A. (1980) The influence of eutrophication on deep lake benthic invertebrate communities. *Progress in Water Technology*, **12**, 161–80.

Scavia, D. (1979) The use of ecological models of lakes in synthesising available information and identifying research needs, in *Perspectives on Lake Ecosystem Modelling* (eds. D. Scavia and A. Robertson), Ann Arbor Science, Ann Arbor, Michigan, pp. 109–68.

Scheffer, M. (1989) Alternative stable states in eutrophic, shallow freshwater systems: a minimal model. *Hydrobiological Bulletin*, **23**, 73–83.

Scheimer, F. (1983) The Parakrama project – scope and objectives, in *Limnology of Parakrama Samudra, Sri Lanka* (ed. F. Scheimer), Dr. W. Junk Publishers, The Hague, pp. 1–17.

Schelske, C.F. and Stoermer, E.F. (1972) Phosphorus, silica and the eutrophication of Lake Michigan, in *Nutrients and Eutrophication: the limiting nutrient controversy* (ed. G.E. Likens), American Society of Limnology and Oceanography, Lawrence, Kansas, pp. 157–71.

Schindler, D.W. (1974) Eutrophication and recovery in experimental lakes: implications for lake management. *Science*, **184**, 897–9.

Schindler, D.W. (1975) Whole lake experiments with phosphorus, nitrogen and carbon. *Verhandlungen Internationale Vereinigung für Theoretische und Angewandte Limnologie*, **19**, 3221–31.

Schindler, D.W. (1976) Biogeochemical evolution of phosphorus limitation in nutrient-enriched lakes of the Precambrian Shield, in *Environmental Biogeochemistry* (ed. J.O. Nriagu), Ann Arbor Science, Ann Arbor, Michigan, pp. 647–64.

Schindler, D.W. (1977) Evolution of phosphorus limitation in lakes. *Science*, **195**, 260–262.

Schindler, D.W. (1978) Factors regulating phytoplankton production and standing crop in the world's freshwaters. *Limnology and Oceanography*, **23**, 478–86.

Schindler, D.W. (1981) Studies of eutrophication in lakes and their relevance to the estuarine environment, in *Estuaries and Nutrients* (eds. B.J. Neilson and L.E. Cronin), Humana Press, Clifton, New Jersey, pp. 71–82.

Schindler, D.W. (1987) Detecting ecosystem responses to anthropogenic stress. *Canadian Journal of Fisheries and Aquatic Sciences*, **44** (Suppl.1), 6–25.

Schindler, D.W. and Fee, E.J. (1974) Experimental Lakes Area: wholelake experiments in eutrophication. *Journal of the Fisheries Research Board of Canada*, **31** 937–53.

Schindler, D.W., Armstrong, F.A.J., Holmgren, S.K. and Brunskill, G.J. (1971) Eutrophication of Lake 227, Experimental Lakes Area, Northwestern Ontario, by addition of phosphate and nitrate. *Journal of the Fisheries Research Board of Canada*, **28**, 1763–82.

Schindler, D.W., Brunskill, G.J., Emersen, S., Broeker, W.S. and Peng, T.H. (1972) Atmospheric carbon dioxide: its role in maintaining phytoplankton standing crops. *Science*, **177**, 1192–4.

Schindler, D.W., Kling, H., Schmidt, R.V., Procopowich, J., Frost, V.E., Reid, R.A. and Capel, M. (1973) Eutrophication of Lake 227 by addition of phosphate and nitrate: the second, third and fourth years of enrichment, 1970, 1971 and 1972. *Journal of the Fisheries Research Board of Canada*, **30**, 1415–40.

Schuman, G.E., Burwell, R.E., Priest, R.F. and Spomer, R.G. (1973) Nitrogen losses in surface runoff from agricultural watersheds on Missouri loess. *Journal of Environmental Quality*, **2**, 299–302.

Scott, R.N. (1975) Studies on some Welsh reservoirs with special reference to Talybont Reservoir, in *The Effects of Storage on Water Quality*, The Water Research Centre, Medmenham, Marlow, pp. 203–38.

Seddon, B. (1972) Aquatic macrophytes as limnological indicators. *Freshwater Biology*, **2**, 107–30.

Seitzinger, S.P. (1988) Denitrification in freshwater and coastal marine ecosystems: ecological and geochemical significance. *Limnology and Oceanography*, **33**, 702–24.

Sephton, D.H. and Harris, G.P. (1984) Physical variability and phytoplankton communities: VI. Day to day changes in primary productivity and species abundance. *Archiv für Hydrobiologie*, **102**, 155–75.

Serruya, C. and Pollinger, U. (1983) *Lakes of the Warm Belt*. Cambridge University Press, Cambridge.

Shapiro, J. (1973) Blue-green algae: why they become dominant. *Science*, **179**, 382–4.

Shapiro, J. (1979) The need for more biology in lake restoration, in *Lake Restoration: Proceedings of a National Conference*, US Environmental Protection Agency, Minneapolis, Minnesota, pp. 161–7.

Shapiro, J. (1990a) Current beliefs regarding dominance by blue-greens: the case for the importance of CO_2 and pH. *Verhandlungen Internationale Vereinigung für Theoretische und Angewandte Limnologie*, **24**, in press.

Shapiro, J. (1990b) Biomanipulation: the next phase – making it stable. *Hydrobiologia*, **200/201**, 13–27.

Shapiro, J. and Wright, D.I. (1984) Lake restoration by biomanipulation: Round Lake Minnesota, the first two years. *Freshwater Biology*, **14**, 371–83.

Shuter, B.J. (1978) Size dependence of phosphorus and nitrogen subsistence quotas in unicellular microorganisms. *Limnology and Oceanography*, **23**, 1248–55.

Simpson, P.S. and Eaton, J.W. (1986) Comparative studies of the photosynthesis of the submerged macrophyte *Elodea canadensis* and the filamentous algae *Cladophora glomerata* and *Spirogyra* sp. *Aquatic Botany*, **24**, 1–12.

Sinker, C.S. (1962) The North Shropshire Meres and Mosses – a background for ecologists. *Field Studies*, **1** 101–7.

Sirenko, L.A. (1980) Toxicity fluctuations and factors determining them. *Developments in Hydrobiology*, **2**, 231–4.

Skulberg, O.M., Codd, G.A. and Carmichael, W.W. (1984) Toxic Blue-green algal blooms in Europe: a growing problem. *Ambio*, **13**, 244–7.

Slater, R.W. and Bangay, G.E. (1979) Action taken to control phosphorus in the Great Lakes, in *Phosphorus Management Strategies for Lakes* (eds. R.C. Loehr, C.S. Martin and W. Rast), Ann Arbor Science, Ann Arbor, pp. 13–26.

Smart, A.S. (1989) *An investigation of the Ecology of Water Distribution Systems*. Ph.D. thesis, University of Leicester.

Smith, I.R. (1974) The structure and physical environment of Loch Leven, Scotland. *Proceedings of the Royal Society of Edinburgh* (B), **74**, 81–100.

Smith, R.V. (1977) Domestic and agricultural contributions to the inputs of phosphorus and nitrogen to Lough Neagh. *Water Research*, **11**, 453–9.

Smith, V.H. (1979) Nutrient dependence of primary productivity in lakes. *Limnology and Oceanography*, **24**, 1051–64.

Smith, V.H. (1982) The nitrogen and phosphorus dependence of algal biomass in lakes: an empirical and theoretical analysis. *Limnology and Oceanography*, **27**, 1101–12.

Smith, V.H. (1983) Low nitrogen to phosphorus ratios favor dominance by blue-green algae in lake phytoplankton. *Science*, **221**, 669–71.

Smyly, W.J.P. (1978) The crustacean zooplankton of Grasmere before and after a change in sewage effluent treatment. *Internationale Revue der Gesampten Hydrobiologie*, **63**, 389–403.

Söderlund, R. and Svensson, B.H. (1976) The global nitrogen cycle, in *Nitrogen, Phosphorus and Sulphur – Global Cycles* (eds. B.H. Svensson and R. Söderlund), Ecological Bulletins 22, Stockholm, pp. 23–73.

Solbé, J.F.l.G. (1987) Water quality, in *Salmon and Trout Farming* (eds. L.M. Laird and T. Needham), Ellis Horwood, Chichester, pp. 69–86.

Sommer, U. (1981) The role of r- and K- selection in the succession of phytoplankton in Lake Constance. *Acta Œcological Œcologia Generalis*, **2**, 327–42.

Söndergaard, M., Riemann, B. and Jörgensen, N.O.G. (1985) Extracellular organic carbon (EOC) released by phytoplankton and bacterial production. *Oikos*, **45**, 323–32.

Southwood, T.R.E. (1966) *Ecological methods*. Chapman and Hall, London.

Spence, D.H.N. (1964) The macrophytic vegetation of freshwater lochs, swamps, and associated fens, in *The vegetation of Scotland* (ed. J.H. Burnett), Oliver and Boyd, Edinburgh, pp. 306–81.

Spence, D.H.N. (1967) Factors controlling the distribution of freshwater macrophytes. *Journal of Ecology*, **55**, 147–70.

Spence, D.H.N. (1975) Light and plant response in fresh water, in *Light as an Ecological Factor* (eds. G.C. Evans, R. Bainbridge and O. Rackham), Blackwells Scientific Publishers, Oxford, pp. 93–134.

Spence, D.H.N. (1982) The zonation of plants in freshwater lakes. *Advances in Ecological Research*, **12**, 37–125.

Sprules, W.G. and Knoechel, R. (1984) Lake ecosystem dynamics based upon functional representations of trophic components in *Trophic Interactions within Aquatic Systems* (eds. D.G. Meyers and J.R. Strickler), American Association for the Advancement of Science, Washington DC, pp. 383–403.

Stansfield, J., Moss, B. and Irvine, K. (1989) The loss of submerged plants with eutrophication III. Potential role of organochlorine pesticides: a paleoecological study. *Freshwater Biology*, **22**, 109–32.

Starling, F.L.R.M. and Rocha, A.J.A. (1990) Experimental study of the impacts of planktivorous fishes on plankton community and eutrophication of a tropical Brazilian reservoir. *Hydrobiologia*, **200/201**, 581–91.

Stauffer, R.E. and Armstrong, D.E. (1986) Cycling of iron, manganese, silica, calcium and potassium in two stratified basins of Shagawa lake, Minnesota. *Geochimica et Cosmochimica Acta*, **50**, 215–29.

Steel, J.A. (1972) The application of fundamental limnological research in water supply system design and management, in *Conservation and Productivity of Natural Waters* (eds. R.W. Edwards and D.J. Garrod), Academic Press, London, pp. 41–68.

Steinberg, C.E.W. and Hartmann, H.M. (1988) Planktonic bloom-forming cyanobacteria and the eutrophication of lakes and rivers. *Freshwater Biology*, **20**, 279–87.

Stewart, W.D.P. (1969) Biological and ecological aspects of nitrogen fixation by free-living organisms. *Proceedings of the Royal Society of London*, B, **172**, 367–88.

Stewart, W.D.P., May, E. and Tuckwell, S.B. (1975) Nitrogen and phosphorus from agricultural land and urbanisation and their fate in shallow freshwater lochs. *Ministry of Agriculture, Fisheries and Food Technical Bulletin*, **32**, 276–305.

Stockner, J.G. (1988) Phototrophic picoplankton: an overview from marine and freshwater ecosystems. *Limnology and Oceanography*, **33**, 765–75.

Stockner, J.G. and Benson, W.W. (1967) The succession of diatom assemblages in recent sediments of Lake Washington. *Limnology and Oceanography*, **12**, 513–32.

Straskraba, M. (1980) The effects of physical variables on freshwater production: analyses based on models, in *The Functioning of Freshwater Ecosystems* (eds E.D. Le Cren and R.H. Lowe-McConnell), Blackwell Scientific Publications, Oxford, pp. 13–84.

Straskraba, M. and Gnauck, A.H. (1985) *Freshwater Ecosystems: Modelling and Simulation*. Elsevier, Amsterdam.

Sullivan, P.F. and Carpenter, S.R. (1982) Evaluation of fourteen trophic state indices for phytoplankton of Indiana lakes and reservoirs. *Environmental Pollution* (A), **27**, 143–53.

Talling, J.F. (1957a) The phytoplankton population as a compound photosynthetic system. *New Phytologist*, **56**, 133–49.

Talling, J.F. (1957b) Photosynthetic characteristics of some freshwater plankton diatoms in relation to underwater radiation. *New Phytologist*, **56**, 29–50.

Talling, J.F. (1966) The annual cycle of stratification and phytoplankton growth in Lake Victoria (East Africa) *Internationale Revue der Gesampten Hydrobiologie*, **51**, 545–621.

Talling, J.F. (1976) The depletion of carbon dioxide from lake water by phytoplankton. *Journal of Ecology*, **64**, 79–121.

Talling, J.F. and Talling, I.B. (1965) The chemical composition of African lake waters. *Internationale Revue der Gesampten Hydrobiologie*, **50**, 421–63.

Taylor, A.H. (1978) An analysis of the trout fishing at Eye Brook – a eutrophic reservoir. *Journal of Animal Ecology*, **47**, 407–23.

Taylor, E.W. (1963) *Report of the Results of the Chemical and Bacteriological Examination of London's Waters*, **41**, 44–6.

Taylor, G.T. (1982) The role of pelagic heterotrophic protozoa in nutrient cycling: a review. *Annales Institute Océanographique, Paris*, **58**, 227–41.

Taylor, M. (1978) Mute Swan census. *Transactions of the Norfolk and Norwich Naturalist Society*, **25**, 7.

Teiling, E. (1916) En Kaledonisk fytoplankton formation. *Svensk Botanisk Tidskrift*, **10**, 506–19.

Tetreault, M.J., Benedict, A.H., Kaempfer, C. and Barth, E.F. (1986) Biological phosphorus removal: a technology evaluation. *Journal of the Water Pollution Control Federation*, **58**, 823–7.

Thienemann, A. (1918) Untersuchungen uber die Beziehungen zwischen dem sauerst offgehalt der Wassers und der Zusammenensetsung der Fauna in norddeutschen See. *Archiv für Hydrobiologie*, **12**, 1–65.

Thomas, A.E. (1969) The progress of eutrophication in Central European Lakes, in *Eutrophication: Causes, Consequences, Correctives* (ed. G.A.

Rohlich), National Academy of Sciences, Washington, DC, pp. 17–28.

Thompson, J.M., Ferguson, A.J.D. and Reynolds, C.S. (1982) Natural filtration rates of zooplankton in a closed system: the derivation of a community grazing index. *Journal of Plankton Research*, **4**, 545–60.

Thornton, J.A. (1987) Aspects of eutrophication management in tropical/ sub-tropical regions. *Journal of the Limnological Society of Southern Africa*, **13**, 25–43.

Thornton, J.A. and Walmsley, R.D. (1982) Applicability of phosphorus budget models to South African man-made lakes. *Hydrobiologia*, **89**, 237–45.

Thornton, J.A., Cochrane, K.L., Jarvis, A.C., Zohary, T., Robarts, R.D. and Chutter, F.M. (1986) An evaluation of management aspects of a hypertrophic African impoundment. *Water Research*, **20**, 413–19.

Threlkeld, S.T. (1988) Planktivory and planktivore biomass effects on zooplankton, phytoplankton and the trophic cascade. *Limnology and Oceanography*, **33**, 1362–75.

Tilman, D. (1977) Resource competition between planktonic algae: an experimental and theoretical approach. *Ecology*, **58**, 338–48.

Tilman, D., Kilham, S.S. and Kilham, P. (1976) Morphometric changes in *Asterionella formosa* colonies under phosphate and silicate limitation. *Limnology and Oceanography*, **21**, 883–4.

Tilman, D., Kiesling, R., Sterner, R., Kilham, S.S. and Johnson, F.A. (1986) Green, blue-green and diatom algae: taxonomic differences in competitive ability for phosphorus, silicon and nitrogen. *Archiv für Hydrobiologie*, **106**, 473–85.

Timms, R.M. and Moss, B. (1984) Prevention of growth of potentially dense phytoplankton populations by zooplankton grazing, in the presence of zooplanktivorous fish, in a shallow wetland ecosystem. *Limnology and Oceanography*, **29**, 472–86.

Titus, J.E., Goldstein, R.A., Adams, M.S., Mankin, J.B., O'Neill, R.V., Weiler, P.R., Shugart, H.H. and Booth, R.S. (1975) A production model for *Myriophyllum spicatum* L. *Ecology*, **56**, 1129–38.

Toerien, D.F. (1975) South African eutrophication problems: a perspective. *Water Pollution Control*, **74**, 134–42.

Toerien, D.F. and Steyn, D.J. (1975) The eutrophication levels of four South African impoundments. *Verhandlungen Internationale Vereinigung für Theoretische und Angewandte Limnologie*, **19**, 1947–56.

Tolland, H.G. (1977) *Destratification/aeration in reservoirs*. Water Research Centre, Medmenham, Technical Report TR 50.

di Toro, D.M., Connor, D.J. and Thomann, R.V. (1975) Phytoplankton-zooplankton-nutrient interaction model for western Lake Erie, in *Systems Analysis*, (ed. B.C. Patten), Academic Press, New York, pp. 423–74.

Townsend, C.R. and Perrow, M.R. (1989) Eutrophication may produce population cycles in roach (*Rutilus rutilus* L.) by two contrasting mechanisms. *Journal of Fish Biology*, **34**, 161–4.

Trifonova, I.S. (1988) Oligotrophic-eutrophic succession of lake phytoplankton, in *Algae and the Aquatic Environment* (ed. F.E. Round), Biopress, Bristol, pp. 107–24.

Troake, R.P., Troake, L.E. and Walling, D.E. (1975) Nitrate loads of south Devon streams. *Ministry of Agriculture, Fisheries and Food Technical Bulletin*, 32, 340–51.

Tundisi, J.G. (1981) Typology of reservoirs in Southern Brazil. *Verhandlungen Internationale Vereinigung für Theoretische und Angewandte Limnologie*, 21, 1031–9.

Twinch, A.J. (1986) The phosphorus status of sediments in a hypereutrophic impoundment (Hartbeespoort Dam): implications for eutrophication management. *Hydrobiologia*, 135, 23–34.

US EPA (1979) *Lake Restoration*. US Environmental Protection Agency, Minneapolis, Minnesota.

Uhl, C. and Jordan, C.F. (1984) Succession and nutrient dynamics following forest cutting and burning in Amazonia. *Ecology*, 65, 1476–90.

Uhlman, D. (1971) Influence of dilution, sinking and grazing rate on phytoplankton populations of hyperfertilised ponds and mocro-ecosystems. *Mitteilungen Internationale Vereinigung Für Theoretische und Angewandte Limnologie*, 19, 100–24.

Usher, M. (1986) *Wildlife Conservation Evaluation*. London, Chapman and Hall.

Utschick, H. (1976) Die Wasservögel als Indikatoren für den ökologischen Zustand von Seen. *Verhandlungen der Ornithologischen Gesellschaft in Bayern*, 22, 395–438.

Uttormark, P.D., Chapin, J.D. and Green, K.M. (1974) *Estimating nutrient loading of lakes from nonpoint sources*. U.S. Environmental Protection Agency, Washington DC, EPA–660/13–74–020.

Verhalen, F.A., Gibbons, H.L.J. and Funk, W.H. (1985) Implications for control of eurasian water milfoil in the Pend Oreille River, in *Lake and Reservoir Management, Practical Applications*, North American Lake Management Society, McAfee, New Jersey, pp. 361–4.

Viner, A.B. (1973) Response of tropical mixed phytoplankton to nutrient enrichments of ammonia and phosphate, and some ecological implications. *Proceedings of the Royal Society of London* (B), 183, 351–70.

Viner, A.B. (1975) The supply of minerals to tropical lakes and rivers (Uganda), in *Coupling of Land and Water Systems* (ed. A.D. Hasler), Springer-Verlag, Berlin.

Viner, A.B. (1977) Relationships of nitrogen and phosphorus to a tropical phytoplankton population. *Hydrobiologia*, 52, 185–96.

Viner, A.B., Breen, C., Golterman, H.L. and Thornton, J.A. (1981) Nutrient budgets, in *The Ecology and Utlisation of African Inland Waters* (eds. J.J. Symoens, M. Burgis and J.J. Gaudet), United Nations Environment Programme, Nairobi, pp. 137–48.

Vollenweider, R.A. (1968) *Water Management Research; Scientific*

Fundamentals of the Eutrophication of Lakes and Flowing Waters, with particular reference to Nitrogen and Phosphorus as factors in Eutrophication. OECD, Paris, Technical Report DAS/CSI/68.27.

Vollenweider, R.A. (1975) Input – Output models with special reference to the phosphorus loading concept in Limnology. *Schweizische Zeitshrift fuer Hydrologie*, 37, 53–84.

Vollenweider, R.A., Munawar, M. and Stadelmann, P. (1974a) A comparative view of phytoplankton and primary production in the Laurentian Great Lakes. *Journal of the Fisheries Research Board of Canada*, 31, 739–62.

Vollenweider, R.A., Talling, J.F. and Westlake, D.F. (1974b) *A Manual on Methods for Measuring Primary Production in Aquatic Environments.* Blackwell Scientific Publishers, Oxford.

Vrhovsek, D., Kosi, G., Kralji, M., Bricelj, M. and Zupan, M. (1985) The effect of restoration measures on the physical, chemical and phytoplankton variables of Lake Bled. *Hydrobiologia*, 127, 219–28.

Watson, S. and Kalff, J. (1981) Relationships between nannoplankton and lake trophic status. *Canadian Journal of Fisheries and Aquatic Science*, 38, 960–7.

Watson, R.A. (1981) *The Limnology of the Thurne Broads.* Ph.D. thesis, University of East Anglia.

Weatherley, A.J. (1972) Growth and Ecology of Fish Populations. Academic Press, London.

Weber, C.A. (1907) Aufbau und Vegetation der Moore Norddeutschlands. *Botanische Jahrbuecher fuer Systematik Pflanzengeschichte und Pflanzengeographie.* (Suppl.) 40, 19–34.

Weibel, S.R. (1969) Urban drainage as a factor in eutrophication, in *Eutrophication: Causes, Consequences, Correctives* (ed. G.A. Rohlich), National Academy of Sciences, Washington, DC, pp. 383–403.

Weiderholm, T. (1976) Chironomids as indicators of water quality in Swedish lakes. *Naturvardsverkets Limnologiska Undersokelser Information*, 10, 1–17.

Welch, E.B. (1980) *Ecological Effects of Waste Water.* Cambridge University Press, Cambridge.

Welch, E.B. (1983) Lake restoration results, in *Lakes and Reservoirs of the World* (ed. F.B. Taub), Elsevier, New York, pp. 557–71.

Welch, E.B. and Patmont, C.R. (1980) Lake restoration by dilution: Moses Lake, Washington. *Water Research*, 14, 1317–25.

Welch, E.B., Spyridakis, D.E., Shuster, J.I. and Horner, R.R. (1986) Declining lake sediment phosphorus release and oxygen deficit following wastewater diversion. *Journal of the Water Pollution Control Federation*, 58, 92–7.

Welcomme, R. (1985) *River Fisheries.* Food and Agriculture Organisation of the United Nations, Rome.

West, W. and West, G.S. (1909) The British freshwater phytoplankton with special reference to the desmid plankton and the distribution of

British desmids. *Proceedings of the Royal Society of London*, **81**, 165–206.

Westlake, D.F. (1975) Macrophytes, in *River Ecology* (ed. B. Whitton), Blackwell Scientific Publications, Oxford, pp. 106–28.

Wetzel, R.G. (1964) A comparative study of the primary productivity of higher aquatic plants, periphyton and phytoplankton in a large, shallow lake. *Internationale Revue der Gesampten Hydrobiologie*, **49**, 1–61.

Wheeler, A. (1978) *Key to the Fishes of Northern Europe*. Frederick Warne, London.

White, E. (1983) Lake eutrophication in New Zealand – a comparison with other countries of the Organisation for Economic Co-operation and Development. *New Zealand Journal of Marine and Freshwater Research*, **17**, 437–44.

Whiteside, M.C. (1983) The mythical concept of eutrophication. *Hydrobiologia*, **103**, 107–11.

Wilkinson, P.D., Bolas, P.M. and Adkins, M.F. (1981) Bewl Bridge Treatment Works. *Journal of the Institution of Water Engineers and Scientists*, **35**, 47–58.

Wilkinson, W.B. and Greene, L.A. (1982) The water industry and the nitrogen cycle. *Philosophical Transactions of the Royal Society of London* (B), **296**, 459–75.

Willemsen, J. (1980) Fishery aspects of eutrophication. *Hydrobiological Bulletin*, **14**, 12–21.

Williams, R.J.B. (1971) The chemical composition of water from land drains at Saxmundham and Woburn, and the influence of rainfall upon nutrient losses. *Report of the Rothamsted Experimental Station for 1970*, **2**, 36–67.

Williams, W.T. and Stephenson, W. (1973) The analyis of three-dimensional data (sites × species × times) in marine ecology. *Journal of Experimental marine Biology and Ecology*, **11**, 207–27.

Wisniewski, R.J. and Planter, M. (1985) Exchange of phosphorus across sediment-water interface (with special attention to the influence of biotic factors) in several lakes of different trophic status. *Verhandlungen Internationale Vereinigung für Theoretische und Angewandte Limnologie*, **22**, 3345–9.

Worthington, E.B. (1975) *The Evolution of IBP*. Cambridge University Press, Cambridge.

Wortley, J.S. (1974) *The Role of Macrophytes in the Ecology of Gastropods and Other Invertebrates in the Norfolk Broads*. Ph.D. thesis, University of East Anglia.

Wortley, J.S. and Phillips, G.L. (1987) Fish mortalities and Prymnesium in the Norfolk Broads, in *Proceedings of the 18th Study Course of the Institute of Fisheries Management, Cambridge* (ed. J.S. Wortley), pp. 152–62.

Wright, D.I. and Shapiro, J. (1984) Nutrient reduction by biomanipulation. *Verhandlungen Internationale Vereinigung für Theoretische und*

Angewandte Limnologie, **22,** 518–24.

Yabro, L.A. (1983) The influence of hydrologic variations on phosphorus cycling and retention in a swamp stream ecosystem, in *The Dynamics of Lotic Ecosystems* (eds. T.D. Fontaine II and S.M. Bartel), Ann Arbor Science, Ann Arbor, Michigan, pp. 223–45.

Youngman, R.E. (1986) Implication for water supply of aerial fertilisation of forests, in *Effects of Land Use on Freshwaters. Agriculture, Forestry, Mineral Exploitation, Urbanisation* (ed. J.F. de L.G. Solbé), Ellis Horwood Ltd, Chichester, pp. 546–8.

Youngman, R.E. (1975) Observations on Farmoor, a eutrophic reservoir in the upper Thames Valley during 1965–73, in *The Effects of Storage on Water Quality,* Water Research Centre, Medmenham, pp. 163–201.

Youngman, R.E. (1986) Implication for water supply of aerial fertilisation of forests, in *Effects of Land Use on Freshwaters. Agriculture, Forestry, Mineral Exploitation, Urbanisation* (ed. J.F. de L.G. Solbé), Ellis Horwood Ltd, Chichester, pp. 546–8.

Index